Mathematical Finance

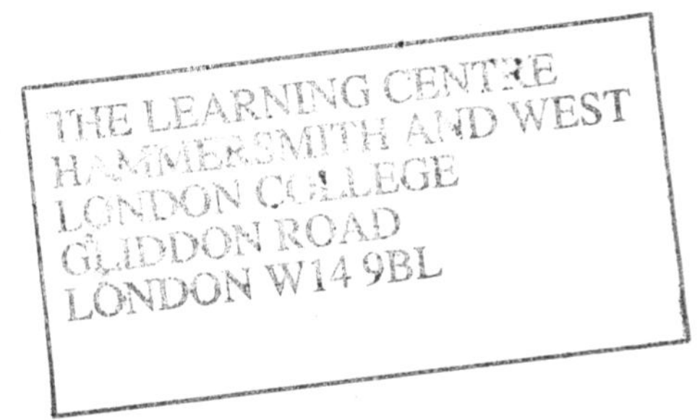

Nikolai Dokuchaev's comprehensive text book provides a systematic, self-sufficient and yet short presentation of the mainstream topics of mathematical finance and related parts of stochastic analysis and statistical finance that covers typical university programs. It can be used as either a point of reference or to provide fundamentals for further research.

Starting with an introduction to probability theory, the book offers a detailed study of discrete and continuous time market models, a comprehensive review of Ito calculus and statistical methods as a basis for statistical estimation of models for pricing, and a detailed discussion of options and their pricing, including American options in continuous time setting. All basic concepts and results are given with proofs and with numerous examples and problems.

This handy introduction to the topic is a useful counterpart to other Routledge books including Barry Goss's *Models of Futures Markets* and *Advanced Mathematical Economics* by Rakesh Vohra. It is suitable for undergraduate and postgraduate courses and advanced degree programs, as well as academics and practitioners.

Nikolai Dokuchaev is Associate Professor in the Department of Mathematics, Trent University, Ontario, Canada.

Routledge advanced texts in economics and finance

Financial Econometrics
Peijie Wang

Macroeconomics for Developing Countries, Second edition
Raghbendra Jha

Advanced Mathematical Economics
Rakesh Vohra

Advanced Econometric Theory
John S. Chipman

Understanding Macroeconomic Theory
John M. Barron, Bradley T. Ewing and Gerald J. Lynch

Regional Economics
Roberta Capello

Mathematical Finance
Core theory, problems and statistical algorithms
Nikolai Dokuchaev

Applied Health Economics
Andrew M. Jones, Nigel Rice, Teresa Bago d'Uva and Silvia Balia

Mathematical Finance

Core theory, problems and statistical algorithms

Nikolai Dokuchaev

LONDON AND NEW YORK

First published 2007 by Routledge
2 Park Square, Milton Park, Abingdon, Oxon OX14 4RN

Simultaneously published in the USA and Canada
by Routledge
270 Madison Ave, New York, NY 10016

Routledge is an imprint of the Taylor & Francis Group, an informa business

Typeset in Times New Roman by Keyword Group Ltd.
Printed and bound in Great Britain by TJ International Ltd, Padstow, Cornwall

British Library Cataloguing in Publication Data
A catalogue record for this book is available from the British Library

Library of Congress Cataloging in Publication Data
A catalog record for this book has been requested

ISBN10: 0-415-41447-4 (hbk)
ISBN10: 0-415-41448-2 (pbk)
ISBN10: 0-203-96472-1 (ebk)

ISBN13: 978-0-415-41447-0 (hbk)
ISBN13: 978-0-415-41448-7 (pbk)
ISBN13: 978-0-203-96472-9 (ebk)

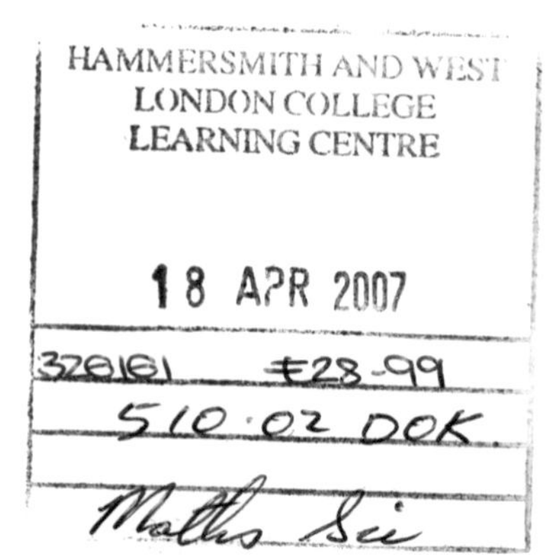

Contents

Preface

Dedicated to Natalia, Lidia, and Mikhail

This book gives a systematic, self-sufficient, and yet short presentation of the mainstream topics of Mathematical Finance and related part of Stochastic Analysis and Statistical Finance that covers typical university programs. It is suitable for undergraduate and graduate students in Mathematics, Statistics, Finance, and Economics. It can be recommended also for academics and practitioners.

The book contains sufficient reference material, including basic models, formulas, and algorithms, plus problems and solutions. A reader who wishes to obtain some basic understanding of the subject may skip proofs and use mathematical theorems as reference. However, there is the second goal of this book: to provide fundamentals for further research and study in this field that has a huge prospectus, and where so many problems are still unsolved. In other words, we want to help readers who intend to do research in Mathematical Finance. For this purpose, it was insufficient to give descriptive material and list the definitions and rules. We provide the fundamentals that give the ability to do research, in particular to create new definitions and rules.

This is why we use the mathematically rigorous way of presentation and give all essential proofs. It is a hard way. To make the ideas more visible, the most technical details of some proofs are omitted, if the main idea of the proof still can be exposed (an example is the proof of the Martingale Representation Theorem in Chapter 4). However, certain angles cannot be cut, and some key proofs cannot be skipped.[1] A reader should be able to set and solve certain theoretical problems after finishing this book. For this purpose, a set of special challenging problems is given for all main topics.[2] The author believes that ability to solve these problems indicates that one is capable of steps toward original research projects.

To keep the course compact, we concentrate on the ultimately important core topics. Moreover, we tried to implement the Okkam razor rule for material selection: we omitted everything that was not necessary for logical and structural

1 In mathematical sciences, understanding of the proofs gives ability to modify them and to extend for new settings. Usually, it is necessary for research.

2 Problems 3.47, 3.70, 5.84, 5.85, 6.25, 6.34, and 9.20, and some others.

completeness. For instance, there are several important approaches to the theory of pricing that are not discussed, including equilibrium concept.

Chapters 1 and 2 give a systematic introduction to probability theory that can be used independently as a reference. Chapter 1 can be also considered as a short introduction to measure and integration theory. Further, we describe generic discrete time market models (Chapter 3) and continuous time market models (Chapter 5). For these two types of models, the book provides all basic concepts of Mathematical Finance (strategies, completeness, risk-neutral pricing, etc.) using numerous examples and problems. Chapter 4 gives a self-sufficient review of Ito calculus that is used for continuous time models. In this chapter, uniqueness for the Ito equation, Kolmogorov's parabolic equations, Martingale representation theorem, and Girsanov theorem, are given with short and relatively simple proofs such that only technical details are omitted.

Chapter 6 addresses binomial trees and American options. Chapter 7 discusses implied volatility. Chapter 8 gives a review of statistical methods. Chapter 9 is devoted to statistical estimation of several generic models for stock prices. In particular, methods of forecast and estimation of the appreciation rates and the volatility are given for log-normal and mean reverting model. Many supporting MATLAB programs are given.

The size of this book is such that it covers approximately three consequent undergraduate or graduate modules such as Mathematical Finance, Statistical Finance, and Stochastic Analysis; for undergraduate modules, some parts can be skipped. Respectively, the set of problems provided is sufficient to cover tutorials for three modules (there are also some solutions). Actually, the book grew out from *Mathematical Finance* and *Statistical Finance* undergraduate modules that the author taught at the University of Limerick, and the problems cover the past exam papers.

Let us list some topics that are *not* covered in this book. Continuous time models with jumps in prices are not addressed: it requires a too-advanced version of stochastic calculus. Optimal portfolio selection is also not covered, with the exception of the solution for the single period market in the simplest version of Markowitz setting (to cover this topic in multi-period or continuous time setting, we need facts from stochastic control theory). Bond market models are described very briefly. The author admits that the corresponding additions could be useful; however, the size of the course in that case would be much bigger.

For further reading, we can recommend the mathematically challenging books of Karatzas and Shreve (1998), Korn (2001), Lambertone and Lapeyre (1996), which cover continuous market models. Discrete time market models were considered by Pliska (1997) and Föllmer and Schied (2002). Shiryaev (1999) considered both discrete time and continuous time models. These books include many advanced results outside of typical university undergraduate and even graduate programs; they can be recommended to readers with some background in stochastic analysis. A more intuitive mathematical approach is offered by Avellaneda (2000), Neftci (1996), Wilmott *et al.* (1997), and Higham (2004). More details for statistical methods can be found in Gujarati (1995) and Söderlind (2005).

Acknowledgments

I wish to thank all my colleagues from the Department of Mathematics and Statistics at the University of Limerick for their support, help, and advice, in establishing the modules as part of the undergraduate program in Financial Mathematics where the topics discussed in this book were taught. I would like to thank all my students for accepting this difficult material with understanding and patience, and for helping to improve it eventually via feedback. I would like to thank Professor A. Rodkina and other colleagues from the University of West Indies, Jamaica, where I had started to work on this project. Finally, I would like to thank Professor B. Goldys, Professor U.G. Haussmann, Professor A.V. Savkin, Professor K.L. Teo, and Professor X.Y. Zhou, for their collaboration dating back several years when I started my career in Mathematical Finance, when this area was completely new to me.

1 Review of probability theory

In probability theory based on Kolmogorov's probability axioms, the model of randomness is the following. It is assumed that there exists a set Ω, and it is assumed that subsets $A \subseteq \Omega$ are random events. Some value $\mathbf{P}(A) \in [0, 1]$ is attached to any event as the probability of an event, and $\mathbf{P}(\Omega) = 1$. To make this model valid, some axioms about possible classes of events are accepted such that the expectation can be interpreted as an integral.

1.1 Measure space and probability space

σ-algebra of events

Let Ω be a non-empty set. We denote by 2^Ω the set of all subsets of Ω.

Example 1.1 Let $\Omega = \{a, b\}$, then $2^\Omega = \{\emptyset, \{a\}, \{b\}, \Omega\}$.

Definition 1.2 *A system of subsets $\mathcal{F} \subset 2^\Omega$ is called an algebra of subsets of Ω if*

(i) $\Omega \in \mathcal{F}$;
(ii) If $A \in \Omega$ then $\Omega \backslash A \in \mathcal{F}$;
(iii) If $A_1, A_2, \ldots, A_n \in \mathcal{F}$, then $\cup_{i=1}^n A_i \in \mathcal{F}$.

Note that (i) and (ii) imply that the empty set $\emptyset$ always belongs to an algebra.

Definition 1.3 *A system of subsets $\mathcal{F} \subset 2^\Omega$ is called a σ-algebra of subsets of Ω if*

(i) It is an algebra of subsets;
(ii) If $A_1, A_2, \ldots \in \mathcal{F}$ (i.e. $\{A_i\}_{i=1}^{+\infty} \subset \mathcal{F}$), then $\cup_{i=1}^{+\infty} A_i \in \mathcal{F}$.

Definition 1.4 *Let Ω be a set, let $\mathcal{F}$ be a σ-algebra of subsets, and let $\mu : \mathcal{F} \to [0, +\infty]$ be a mapping.*

(i) We said that μ is a σ-additive measure if $\mu(\cup_{i=1}^{+\infty} A_i) = \sum_{i=1}^{+\infty} \mu(A_i)$ for any $A_1, A_2, \ldots \in \mathcal{F}$ such that $A_i \cap A_j = \emptyset$ if $i \neq j$. In that case, the triplet $(\Omega, \mathcal{F}, \mu)$ is said to be a measure space.
(ii) If $\mu(\Omega) < +\infty$, then the measure μ is said to be finite.
(iii) If $\mu(\Omega) = 1$, then the measure μ is said to be a probability measure.

To make notations more visible, we shall use the symbol $\mathbf{P}$ for the probability measures.

Definition 1.5 *Consider a measure space $(\Omega, \mathcal{F}, \mu)$. Assume that some property holds for all $\omega \in \Omega_1$, where $\Omega_1 \in \mathcal{F}$ is such that $\mu(\Omega \backslash \Omega_1) = 0$. We say that this property holds a.e. (almost everywhere). In the case of a probability measure, we say that this property holds with probability 1, or a.s. (almost surely).*

In probability theory based on Kolmogorov's probability axioms, the following definition is accepted.

Definition 1.6 *A measure space $(\Omega, \mathcal{F}, \mathbf{P})$ is said to be a probability space if $\mathbf{P}$ is a probability measure, i.e., $\mathbf{P}(\Omega) = 1$. Elements $\omega \in \Omega$ are said to be elementary events, and sets $A \in \mathcal{F}$ are said to be events (or random events). Correspondingly, $\mathcal{F}$ is the σ-algebra of events.*

Under these axioms, $A \cap B$ means the event 'A and B' (or $A \cdot B$), and $A \cup B$ means the event 'A or B' (or $A + B$), where A and B are events.

A random event $A = \{\omega\}$ is a set of elementary events.

Example 1.7 For $\Omega = [0, 1]$, a probability measure may be defined such that

$$\mathbf{P}((a, b]) = \mathbf{P}([a, b)) = \mathbf{P}((a, b)) = \mathbf{P}([a, b]) = b - a$$

for all intervals, where $0 \leq a < b \leq 1$. Clearly, the set of all intervals does not form an algebra and, therefore, it does not form a σ-algebra. The question arises for which σ-algebra this measure can be defined. A natural solution is to take the minimal σ-algebra that contains all open intervals. It is the so-called Borel σ-algebra discussed below. It can be shown that this σ-algebra contains all closed and semi-open intervals as well.

Example 1.8 For $\Omega = \mathbf{R}$, a probability measure may be defined such that

$$\mathbf{P}((a, b]) = \mathbf{P}([a, b)) = \mathbf{P}((a, b)) = \mathbf{P}([a, b]) = \int_a^b p(x)dx,$$

where $p : \mathbf{R} \to \mathbf{R}$ is a function such that

$$p(x) \geq 0, \qquad \int_{\mathbf{R}} p(x)dx = 1.$$

Again, this measure can be defined on the minimal σ-algebra that contains all open intervals (Borel σ-algebra).

Completeness

Definition 1.9 *A σ-algebra $\mathcal{F}$ is said to be complete (with respect to a measure $\mu : \mathcal{F} \to \mathbf{R}$) if the following is satisfied: if $A \in \mathcal{F}$, $B \subset A$, and $\mu(A) = 0$, then $B \in \mathcal{F}$.*

Problem 1.10 Prove that $\mu(B) = 0$ under the assumptions of the definition above.

Definition 1.11 *Let $(\Omega, \mathcal{F}, \mu)$ be a measure space. Let $\bar{\mathcal{F}}$ be a minimal σ-algebra such that $\mathcal{F} \subseteq \bar{\mathcal{F}}$ and $\bar{\mathcal{F}}$ is complete with respect to μ. Then $\bar{\mathcal{F}}$ is called the completion with respect to μ (in the literature, it is sometimes called the μ-augmentation of $\mathcal{F}$).*

Remark 1.12 The completion of the Borel σ-algebra defined in Example 1.7 is called the Lebesgue σ-algebra.

We shall consider complete probability spaces only.

1.2 Random variables

Notation: Let X and Y be two sets, let $f : X \to Y$ be a mapping, and let $B \subset Y$. We denote $f^{-1}(B) \triangleq \{x \in X : f(x) \in B\}$. *(Note that we do not exclude the case when the inverse function $f^{-1} : Y \to X$ does not exist.)*

Definition 1.13 *(i) Let $(\Omega, \mathcal{F}, \mu)$ be a complete measure space. A mapping $\xi : \Omega \to \mathbf{R}$ is said to be measurable (with respect to $\mathcal{F}$), if $\xi^{-1}(D) \in \mathcal{F}$ for any open set $D \subset \mathbf{R}$.*

(ii) Let $(\Omega, \mathcal{F}, \mathbf{P})$ be a probability space. A measurable mapping $\xi : \Omega \to \mathbf{R}$ is said to be a random variable (on this probability space).

As can be seen from the definitions, a mapping may be a random variable for some $\mathcal{F}$ and not be a random variable for some different $\mathcal{F}$.

Notation: *iff* means if and only if.

Proposition 1.14 *A mapping $\xi : \Omega \to \mathbf{R}$ is a random variable iff $\xi^{-1}(I) \in \mathcal{F}$ for any semi-open interval $I = (a, b] \subset \mathbf{R}$.*

Problem 1.15 Introduce and describe a probability space with four elementary events only: $\Omega = \{a, b, c, d\}$. Suggest at least three different σ-algebras of events and measures. Describe random variables $\xi \colon \Omega \to \mathbf{R}$. Give an example of non-measurable function $\xi : \Omega \to \mathbf{R}$.

Problem 1.16 Introduce and describe a probability space with $\Omega = [0, 1]$ and with a σ-algebra of events that contains only four events. Describe measurable functions $\xi : \Omega \to \mathbf{R}$. Give an example of non-measurable function.

Definition 1.17 *(i) Let $(\Omega, \mathcal{F}, \mu)$ be a complete measure space. A mapping $\xi : \Omega \to \mathbf{R}^n$ is said to be measurable (with respect to $\mathcal{F}$), if $\xi^{-1}(D) \subset \mathcal{F}$ for any open set $D \subseteq \mathbf{R}^n$.*

(ii) Let $(\Omega, \mathcal{F}, \mathbf{P})$ be a probability space. A measurable mapping $\xi : \Omega \to \mathbf{R}^n$ is said to be a random vector.

Note that $\xi = (\xi_1, \dots, \xi_n)$ is a random vector iff all components ξ_i are random variables.

Definition 1.18 *Two random vectors ξ_1 and ξ_2 are said to be* **P***-indistinguishable (or* **P***-equivalent, or equivalent) if $\mathbf{P}(\xi_1 \neq \xi_2) = 0$. In that case, ξ_1 is said to be a modification of ξ_2.*

1.3 Expectations

Let $(\Omega, \mathcal{F}, \mathbf{P})$ be a probability space.

Definition 1.19 *(i) A random variable $\xi : \Omega \to \mathbf{R}$ is said to be finitely valued if*

$$\exists n \in \mathbf{N},\ c_1, \dots, c_n \in \mathbf{R},$$

$$A_1, A_2, \dots, A_n \in \mathcal{F} : \xi(\omega)|_{\omega \in A_i} \equiv c_i, \quad \xi(\omega)|_{\omega \in \Omega \setminus \cup_i A_i} \equiv 0.$$

(ii) The value

$$\sum_{i=n} c_i \mathbf{P}(A_i),$$

is said to be the integral $\int_\Omega \xi(\omega)\mathbf{P}(d\omega)$ (i.e., the integral of ξ over Ω with respect to the measure **P***). It is also said to be the expectation $\mathbf{E}\xi$ of ξ (or the expected value, or the mathematical expectation, or the mean).*

Let $\mathbb{I}_A(\omega)$ denote the indicator function of a set A:

$$\mathbb{I}_A(\omega) \stackrel{\Delta}{=} \begin{cases} 1, & \omega \in A \\ 0, & \omega \notin A \end{cases}$$

Example 1.20 If $A \in \mathcal{F}$, then the function $\mathbb{I}_A : \Omega \to \mathbf{R}$ is measurable and finitely valued, and

$$\mathbf{E}\mathbb{I}_A = \int_\Omega \mathbb{I}_A(\omega)\mathbf{P}(d\omega) = \mathbf{P}(A).$$

Note that $\int_\Omega \mathbb{I}_A(\omega)\xi(\omega)\mathbf{P}(d\omega)$ can also be written as $\int_A \xi(\omega)\mathbf{P}(d\omega)$.

Problem 1.21 Give an example of $(\Omega, \mathcal{F}, \mathbf{P})$ and of a finitely valued measurable function. Find its expectation.

Definition 1.22[1] *A non-negative random variable $\xi : \Omega \to \mathbf{R}$ is said to be integrable if there exists a non-decreasing sequence of non-negative finitely valued random variables $\xi_i(\cdot)$ such that $\xi_i(\omega) \to \xi(\omega)$ as $i \to +\infty$ a.s. (almost surely) (i.e., with probability 1), and such that*

$$\sup_n \int_\Omega \xi_n(\omega)\mathbf{P}(d\omega)$$

is finite.

Lemma 1.24 *Under the assumptions of the previous definition, there exists the limit of $\mathbf{E}\xi_n = \int_\Omega \xi_n(\omega)\mathbf{P}(d\omega)$ as $n \to +\infty$. This limit is uniquely defined (i.e., it does not depend on the choice of $\{\xi_k\}$), and this limit is said to be the integral $\int_\Omega \xi(\omega)\mathbf{P}(d\omega)$, or the expectation $\mathbf{E}\xi$.*

Notation: We denote $x^+ \triangleq \max(x, 0)$ and $x^- = \max(-x, 0)$.

Definition 1.25 *A random variable $\xi : \Omega \to \mathbf{R}$ is said to be integrable if ξ^+ and ξ^- are integrable (note that $\xi = \xi^+ - \xi^-$). In that case, the value*

$$\mathbf{E}\xi^+ - \mathbf{E}\xi^- = \int_\Omega \xi^+(\omega)\mathbf{P}(d\omega) - \int_\Omega \xi^-(\omega)\mathbf{P}(d\omega)$$

is said to be $\mathbf{E}\xi = \int_\Omega \xi(\omega)\mathbf{P}(d\omega)$.

Resuming, we may say that, for a probability space $(\Omega, \mathcal{F}, \mathbf{P})$, a measurable (with respect to $\mathcal{F}$) function $\xi : \Omega \to \mathbf{R}$ is a random variable, and the integral $\mathbf{E}\xi = \int_\Omega \xi(\omega)\mathbf{P}(d\omega)$ is the expectation (or mathematical expectation, or *mean*).

1 There is an equivalent definition.

Definition 1.23 *A random variable $\xi : \Omega \to \mathbf{R}$ is said to be integrable if there exists a sequence of finitely valued random variables ξ_i such that $\xi_i(\omega) \to \xi(\omega)$ as $i \to +\infty$ a.s., and such that*

$$\lim_{n,m\to+\infty} \int_\Omega |\xi_n(\omega) - \xi_m(\omega)|\mathbf{P}(d\omega) = 0.$$

Notation: Let $p \in [1, +\infty)$. We denote by $\mathcal{L}_p(\Omega, \mathcal{F}, \mathbf{P})$ the set of all random variables ξ on a probability space $(\Omega, \mathcal{F}, \mathbf{P})$ such that $\mathbf{E}|\xi|^p < +\infty$. In addition, we denote by $\mathcal{L}_\infty(\Omega, \mathcal{F}, \mathbf{P})$ the set of all random variables ξ on a probability space $(\Omega, \mathcal{F}, \mathbf{P})$ such that there exists a (non-random) constant $c = c(\xi) > 0$ such that $|\xi| \le c$ a.s.

With these notations, the set of all integrable random variables is $\mathcal{L}_1(\Omega, \mathcal{F}, \mathbf{P})$.

If $\xi \in \mathcal{L}_2(\Omega, \mathcal{F}, \mathbf{P})$, then $\mathbf{E}\xi^2 < +\infty$. In that case, the variance of ξ is defined: $\mathrm{Var}\,\xi \triangleq \mathbf{E}\xi^2 - (\mathbf{E}\xi)^2 = \mathbf{E}(\xi - \mathbf{E}\xi)^2$.

Note that it can happen that $\xi, \eta \in \mathcal{L}_p(\Omega, \mathcal{F}, \mathbf{P})$, $\xi \ne \eta$, and $\mathbf{P}(\xi \ne \eta) = 0$ (in other words, they are **P**-indistinguishable). Formally, ξ and η are different elements of $\mathcal{L}_p(\Omega, \mathcal{F}, \mathbf{P})$. This can be inconvenient, so we introduce the following notation.

Notation: For $p \in [1, +\infty]$, we denote by $L_p(\Omega, \mathcal{F}, \mathbf{P})$ the set of classes of random variables from $\mathcal{L}_p(\Omega, \mathcal{F}, \mathbf{P})$ that are **P**-equivalent. In other words, if $\mathbf{P}(\xi \ne \eta) = 0$, then $\xi = \eta$, meaning that they represent the same element of $L_p(\Omega, \mathcal{F}, \mathbf{P})$, i.e., they are in the same class of equivalency.

About $\mathcal{L}_p$ and L_p as linear spaces

The current course is constructed such that we do not need to refer to the definition of linear normed spaces, inner product spaces, and their properties that are usually studied in Functional Analysis or Function Spaces courses. However, it may be useful for readers who are familiar with these definitions to note that $\mathcal{L}_p(\Omega, \mathcal{F}, \mathbf{P})$ and $L_p(\Omega, \mathcal{F}, \mathbf{P})$ are linear spaces. In addition, $L_p(\Omega, \mathcal{F}, \mathbf{P})$ is a metric space, a linear normed space, and a Banach space for any $p \ge 1$, and $L_2(\Omega, \mathcal{F}, \mathbf{P})$ is an inner product space and a Hilbert space.

Remark 1.26 (i) Any bounded random variable is integrable.

(ii) The integrability is defined by the probability distribution of the random variable (the distributions are discussed below).

(iii) A random variable is integrable iff $|\xi|$ is integrable.

Theorem 1.27 *Let ξ, η be integrable ransom variables, $\alpha \in \mathbf{R}$. Then*

(i) $\mathbf{E}(\xi + \eta) = \mathbf{E}\xi + \mathbf{E}\eta$, $\mathbf{E}(\alpha\xi) = \alpha\mathbf{E}\xi$;

(ii) If $\mathbf{P}(\xi \ne \eta) = 0$, *then* $\mathbf{E}\xi = \mathbf{E}\eta$.

Proof follows from the construction of an appropriate sequence of finitely valued functions. □

We say that a random variable ξ has the probability distribution $N(a, \sigma^2)$ and write $\xi \sim N(a, \sigma^2)$ if ξ is a Gaussian random variable such that $\mathbf{E}\xi = a$ and $\mathrm{Var}\,\xi = \sigma^2$ (see Section 1.8 below).

Example 1.28 It is possible that a random variable is non-integrable. For instance, if $\xi \sim N(0, \sigma^2)$, then e^{ξ^2} is integrable for some $\sigma > 0$ and it is non-integrable for some $\sigma > 0$.

Problem 1.29 Find when e^{ξ^2} is integrable for $\xi \sim N(0, \sigma^2)$.

Definition 1.30 *Let ξ be a random variable that is non-integrable.*

(i) Let $\xi \geq 0$. We may say that the expectation of ξ does not exist; alternatively, we may say that $\mathbf{E}\xi = +\infty$.
(ii) Let $\xi \leq 0$. We may say that the expectation of ξ does not exist; alternatively, we may say that $\mathbf{E}\xi = -\infty$.
(iii) Let ξ^- be integrable. It follows that ξ^+ is non-integrable (since we have assumed that ξ is non-integrable). We may say that $\mathbf{E}\xi = +\infty$.
(iv) Let ξ^+ be integrable, then ξ^- is non-integrable, and we may say that $\mathbf{E}\xi = -\infty$.

Remark 1.31 In fact, the definitions above give a brief description of the measure theory and integration theory, which covers the theory of Lebesgue's integral.

1.4 Equivalent probability measures

Definition 1.32 *Let $(\Omega, \mathcal{F}, \mathbf{P}_i)$ be two probability spaces with the same $(\Omega, \mathcal{F})$ and with different $\mathbf{P}_i$, $i = 1, 2$. The measures $\mathbf{P}_i$ are said to be equivalent if they have the same sets of zero sets, i.e.,*

$$\mathbf{P}_1(A) = 0 \Leftrightarrow \mathbf{P}_2(A) = 0, \quad A \in \mathcal{F}.$$

Theorem 1.33 *(Radon–Nikodim theorem). The measures $\mathbf{P}_1$ and $\mathbf{P}_2$ are equivalent iff there exist $Z \in \mathcal{L}_1(\Omega, \mathcal{F}, \mathbf{P}_2)$ such that $Z(\omega) > 0$ a.s., and*

$$\mathbf{P}_1(A) = \int_A Z(\omega)\mathbf{P}_2(d\omega) \quad \forall A \in \mathcal{F}.$$

We say that Z is the Radon–Nikodim derivative: $Z = d\mathbf{P}_1/d\mathbf{P}_2$.

In that case, $\mathbf{E}_1\xi = \mathbf{E}_2 Z\xi$ for any $\mathbf{P}_1$-integrable random variable ξ, where $\mathbf{E}_i$ is the expectation under the measure $\mathbf{P}_i$.

1.5 Conditional probability and expectation

Definition 1.34 *Let A, B be random events. The conditional probability $\mathbf{P}(A \mid B)$ is defined as $\mathbf{P}(A \mid B) \triangleq \mathbf{P}(A \cdot B)/\mathbf{P}(B)$. (In fact, it is the probability of A under the condition that the event B occurs.)*

Remember that $\mathcal{L}_2(\Omega, \mathcal{F}, \mathbf{P})$ denotes the set of all random variables ξ on a probability space $(\Omega, \mathcal{F}, \mathbf{P})$ such that $\mathbf{E}\xi^2 < +\infty$.

Proposition 1.35 *Let* $\xi \in \mathcal{L}_2(\Omega, \mathcal{F}, \mathbf{P})$. *Then*

$$\mathbf{E}|\xi - \mathbf{E}\xi|^2 \le \mathbf{E}|\xi - c|^2 \quad \forall c \in \mathbf{R}.$$

Proof. Clearly,

$$\begin{aligned}\mathbf{E}(\xi - c)^2 &= \mathbf{E}([\xi - \mathbf{E}\xi] + [\mathbf{E}\xi - c])^2 \\ &= \mathbf{E}(\xi - \mathbf{E}\xi)^2 + 2c\mathbf{E}(\xi - \mathbf{E}\xi) + (\mathbf{E}\xi - c)^2 \\ &= \mathbf{E}(\xi - \mathbf{E}\xi)^2 + (\mathbf{E}\xi - c)^2 \ge \mathbf{E}(\xi - \mathbf{E}\xi)^2. \quad \square\end{aligned}$$

Proposition 1.35 states that $c = \mathbf{E}\xi$ is the solution of the minimization problem

$$\text{Minimize} \quad \mathbf{E}(\xi - c)^p \quad \text{over} \quad c \in \mathbf{R}$$

with $p = 2$. (If $p \neq 2$, then it is not true for the general case.) This fact helps to justify the following definition.

Definition 1.36 *Let* $\xi \in \mathcal{L}_2(\Omega, \mathcal{F}, \mathbf{P})$. *Let* $\mathcal{G}$ *be a* σ*-algebra such that* $\mathcal{G} \subseteq \mathcal{F}$. *A random variable* $\mathbf{E}\{\xi \mid \mathcal{G}\}$ *from* $\mathcal{L}_2(\Omega, \mathcal{G}, \mathbf{P})$ *such that*

$$\mathbf{E}|\xi - \mathbf{E}\{\xi \mid \mathcal{G}\}|^2 \le \mathbf{E}|\xi - \eta|^2 \quad \forall \eta \in \mathcal{L}_2(\Omega, \mathcal{G}, \mathbf{P})$$

is called the conditional expectation.

(Note that $(\Omega, \mathcal{G}, \mathbf{P})$ is also a probability space.)

In addition to the formal definition, it may be useful to keep in mind the following intuitive description: the conditional expectation $\mathbf{E}\{\xi \mid \mathcal{G}\}$ is the expectation of a random variable ξ in an imaginary universe, where an observer knows about all events from $\mathcal{G}$, if they occur or not.

Theorem 1.37 *Let a* σ*-algebra* $\mathcal{G}$ *and a random vector* ξ *be given. Then:*

(i) The conditional expectation $\mathbf{E}\{\xi \mid \mathcal{G}\}$ *is uniquely defined (up to* $\mathbf{P}$*-equivalency; i.e., all versions of* $\mathbf{E}\{\xi \mid \mathcal{G}\}$ *are* $\mathbf{P}$*-indistinguishable);*
(ii) $\mathbf{E}(\xi - \mathbf{E}\{\xi \mid \mathcal{G}\})\eta = 0$ *for all* $\eta \in \mathcal{L}_2(\Omega, \mathcal{G}, \mathbf{P})$*;*
(iii) $\mathbf{E}\xi = \mathbf{E}\,\mathbf{E}\{\xi \mid \mathcal{G}\}$*;*
(iv) Let $\mathcal{G}_0$ *be a* σ*-algebra such that* $\mathcal{G}_0 \subseteq \mathcal{G}$. *Then* $\mathbf{E}\{\xi \mid \mathcal{G}_0\} = \mathbf{E}\{\mathbf{E}\{\xi \mid \mathcal{G}\}|\mathcal{G}_0\}$.

By this theorem, the conditional expectation may be interpreted as a projection.[2]

Example 1.38 If $\mathcal{G} = \{\emptyset, \Omega\}$ (the trivial σ-algebra), then $\mathbf{E}\{\xi \mid \mathcal{G}\} = \mathbf{E}\xi$.

Definition 1.39 *The conditional probability measure* $\mathbf{P}\{\cdot \mid \mathcal{G}\} : \mathcal{F} \to [0,1]$ *is defined as*

$$\mathbf{P}(A \mid \mathcal{G}) \triangleq \mathbf{E}\{\mathbb{I}_A \mid \mathcal{G}\}.$$

It can be shown that it is a probability measure, and that $(\Omega, \mathcal{F}, \mathbf{P}(\cdot \mid \mathcal{G}))$ is a probability space.

Problem 1.40 Let $\Omega = \{\omega_1, \omega_2, \omega_3\}$, $\mathcal{F} = 2^\Omega$, $\mathbf{P}(\{\omega_i\}) = 1/3$, $i = 1, 2, 3$. Let $\xi(\omega_i) = i$. Let $\mathcal{G} = \{\emptyset, \Omega, \{\omega_1\}, \{\omega_2, \omega_3\}\}$.
(i) Prove that $\mathcal{G}$ is a σ-algebra. (ii) Find $\mathbf{P}|_{\mathcal{G}}$. (iii) Find $\mathbf{E}\{\xi \mid \mathcal{G}\}$.

In fact, $(\Omega, \mathcal{F}, \mathbf{P}(\cdot \mid \mathcal{G}))$ is a probability space, where the probabilities are calculated by an observer who knows whether any event from $\mathcal{G}$ occurs or not. We illustrate below this statement via σ-algebras generated by a random vector.

1.6 The σ-algebra generated by a random vector

Definition 1.41 *Borel σ-algebra of subsets in $\mathbf{R}^n$ is the minimal σ-algebra that contains all sets $\{x \in \mathbf{R}^n : a_i < x_i \leq b_i\}$ (In fact, it is also the minimal σ-algebra that contains all open sets.) A function $f : \mathbf{R}^n \to \mathbf{R}$ is said to be measurable (or Borel measurable) if it is measurable with respect to this σ-algebra.*

Definition 1.42 *Let $\xi : \Omega \to \mathbf{R}$ be a random variable. Let $\mathcal{F}_\xi$ be the minimal σ-algebra such that $\mathcal{F}_\xi$ includes all random events $\{a < \xi < b\}$, where $a, b \in \mathbf{R}$. We say that $\mathcal{F}_\xi$ is the σ-algebra generated by ξ.*

This definition can be extended for a vector case.

Definition 1.43 *Let $\xi : \Omega \to \mathbf{R}^n$ be a random vector (i.e., all its components are random variables). Let $\mathcal{F}_\xi$ be the minimal σ-algebra such that $\mathcal{F}_\xi$ includes all events $\{\xi \in B\}$ for all Borel sets B in $\mathbf{R}^n$ (or for all open sets). We say that $\mathcal{F}_\xi$ is the σ-algebra generated by ξ.*

2 Usually, we do not refer to the theory of Hilbert spaces. However, it may be useful to keep in mind that $L_2(\Omega, \mathcal{F}, \mathbf{P})$ is a Hilbert with the inner product (scalar product) $\langle \xi, \eta \rangle \triangleq \mathbf{E}\xi\eta$. Then $L_2(\Omega, \mathcal{G}, \mathbf{P})$ is a linear subspace of $L_2(\Omega, \mathcal{F}, \mathbf{P})$, and $\mathbf{E}\{\xi \mid \mathcal{G}\}$ is the projection of ξ on this subspace. As usual, statement (ii) of Theorem 1.37 above means that $\xi - \mathbf{E}\{\xi \mid \mathcal{G}\} \perp \eta$ for all $\eta \in L_2(\Omega, \mathcal{G}, \mathbf{P})$; here $\perp$ means the orthogonality, i.e., $\xi \perp \eta$ means that $\langle \xi, \eta \rangle \triangleq \mathbf{E}\xi\eta = 0$.

Notation: (i) $\sigma(\xi)$ denotes the σ-algebra generated by ξ; (ii) $\bar{\sigma}(\xi)$ denotes the completion of the σ-algebra generated by ξ. (Sometimes we use other notations, for instance, $\mathcal{F}_\xi$.)

Definition 1.44 *Let ξ and η be random variables. Then* $\mathbf{E}\{\eta \mid \xi\} \triangleq \mathbf{E}\{\eta \mid \mathcal{F}_\xi\}$.

In fact, the σ-algebra represents the set of all random events generated by ξ, and $\mathbf{P}(\cdot \mid \mathcal{F}_\xi)$ is the modification of the original probability $\mathbf{P}$ for an observer for whom ξ is known.

Remember that $\mathbf{E}\{\eta \mid \mathcal{F}_\xi\}$ is the best (in mean variance sense) estimate of η obtained via observations of ξ (see Definition 1.36).

Theorem 1.45 *Let $\xi : \Omega \to \mathbf{R}^n$ be a random vector. Let η be a $\mathcal{F}_\xi$-measurable random variable. Then there exists a (non-random) function $f : \mathbf{R}^n \to \mathbf{R}$ such that*[3] $\eta = f(\xi)$.

Note that any η generates its own f in the previous theorem.

Corollary 1.46 *If $\zeta \in \mathcal{L}_2(\Omega, \mathcal{F}, \mathbf{P})$, then there exists a function $f : \mathbf{R}^n \to \mathbf{R}$ such that* $\mathbf{E}\{\zeta \mid \xi\} = \mathbf{E}\{\zeta \mid \mathcal{F}_\xi\} = f(\xi)$.

Problem 1.47 Let $(\Omega, \mathcal{F}, \mathbf{P})$ be a probability space with $\Omega = \{\omega_1, \omega_2, \omega_3\}$, $\mathcal{F} = 2^\Omega$,

$$\mathbf{P}(\{\omega_1\}) = 1/4, \quad \mathbf{P}(\{\omega_2\}) = 1/4, \quad \mathbf{P}(\{\omega_3\}) = 1/2.$$

Let $\xi(\omega_i) = i - 2$. Let $\eta = |\xi|$. Find $\mathbf{E}\{\xi \mid \eta\}$. Express $\mathbf{E}\{\xi \mid \eta\}$ as a deterministic function of η.

Solution. We have that

$$\eta(\omega) = \begin{cases} 0 & \omega = \omega_2 \\ 1 & \omega = \omega_1 \text{ or } \omega = \omega_3 \end{cases}$$

Hence η generates the σ-algebra $\mathcal{F}_\eta = \{\Omega, \emptyset, \{\omega_2\}, \{\omega_1, \omega_3\}\}$. By the definition, $\mathbf{E}\{\xi \mid \eta\} = \mathbf{E}\{\xi \mid \mathcal{F}_\eta\} = \zeta$, where ζ is an $\mathcal{F}_\eta$-measurable random variable such that $\mathbf{E}(\zeta - \xi)^2$ is minimal over all $\mathcal{F}_\eta$-measurable random variables. An $\mathcal{F}_\eta$-measurable random variable has a form

$$\zeta(\omega) = \begin{cases} \alpha & \omega = \omega_2 \\ \beta & \omega = \omega_1 \text{ or } \omega = \omega_3 \end{cases}, \qquad \alpha, \beta \in \mathbf{R}.$$

3 f is also measurable.

Let $f(\alpha,\beta) = \mathbf{E}(\zeta - \xi)^2$. It suffices to find (α, β) such that $f(\alpha,\beta) = \mathbf{E}(\zeta - \xi)^2 =$ min. We have that

$$f(\alpha,\beta) = \mathbf{E}(\zeta-\xi)^2 = \mathbf{P}(\{\omega_1\})(\beta+1)^2 + \mathbf{P}(\{\omega_2\})(\alpha-0)^2 + \mathbf{P}(\{\omega_3\})(\beta-1)^2$$

$$= \tfrac{1}{4}(\beta+1)^2 + \tfrac{1}{4}\alpha^2 + \tfrac{1}{2}(\beta-1)^2,$$

and

$$\frac{\partial f}{\partial \alpha}(\alpha,\beta) = \frac{\alpha}{2}, \quad \frac{\partial f}{\partial \beta}(\alpha,\beta) = \frac{1}{2}(\beta+1) + \beta - 1 = \frac{3}{2}\beta - \frac{1}{2}.$$

Hence the only minimum of f is at $(\alpha, \beta) = (0, 1/3)$. Therefore,

$$\mathbf{E}\{\xi \mid \eta\}(\omega) = \begin{cases} 0 & \omega = \omega_2 \\ 1/3 & \omega = \omega_1 \text{ or } \omega = \omega_3. \end{cases}$$

Clearly, $\mathbf{E}\{\xi \mid \eta\}$ can be expressed as a deterministic function of η:

$$\mathbf{E}\{\xi \mid \eta\} = f(\eta), \quad \text{where} \quad f(x) = \begin{cases} 0 & x = 0 \\ 1/3 & x \neq 0. \end{cases} \qquad \square$$

1.7 Independence

Definition 1.48 *Two random events A and B are said to be independent iff $\mathbf{P}(A \cdot B) = \mathbf{P}(A)\mathbf{P}(B)$.*

Note that if A and B are independent events then $\mathbf{P}(A \mid B) = \mathbf{P}(A)$.

Definition 1.49 *Two σ-algebras of events $\mathcal{F}_1$ and $\mathcal{F}_2$ are said to be independent if any events $A \in \mathcal{F}_1$ and $B \in \mathcal{F}_2$ are independent. Otherwise, these σ-algebras are said to be dependent.*

Definition 1.50 *Two random vectors $\xi : \Omega \to \mathbf{R}^n$ and $\eta : \Omega \to \mathbf{R}^m$ are said to be independent if the σ-algebras $\mathcal{F}_\xi$ and $\mathcal{F}_\eta$ (generated by ξ and η respectively) are independent. Otherwise, these random vectors are said to be dependent.*

Theorem 1.51 *The following statements are equivalent:*

- *Two random vectors $\xi : \Omega \to \mathbf{R}^n$ and $\eta : \Omega \to \mathbf{R}^m$ are independent;*
- $\mathbf{P}((\xi,\eta) \in D_1 \times D_2) = \mathbf{P}(\xi \in D_1)\mathbf{P}(\eta \in D_2)$ *for any open set $D_1 \subset \mathbf{R}^n$, $D_2 \subset \mathbf{R}^m$;*
- $\mathbf{P}((\xi,\eta) \in D_1 \times D_2) = \mathbf{P}(\xi \in D_1)\mathbf{P}(\eta \in D_2)$ *for any Borel set $D_1 \subset \mathbf{R}^n$, $D_2 \subset \mathbf{R}^m$;*
- $\mathbf{E}f(\xi)g(\eta) = \mathbf{E}f(\xi)\mathbf{E}g(\eta)$ *for all functions $f : \mathbf{R}^n \to \mathbf{R}$, $g : \mathbf{R}^m \to \mathbf{R}$ such that the corresponding expectations are well defined;*
- $\mathbf{E}e^{i\lambda\xi + i\mu\eta} = \mathbf{E}e^{i\lambda\xi}\mathbf{E}e^{i\mu\eta}$ *for all $\lambda, \mu \in \mathbf{R}$, where $i = \sqrt{-1}$;*

- $\mathbf{P}(\xi \in D \mid \eta) = \mathbf{P}(\xi \in D)$ *for any open set* $D \subset \mathbf{R}^n$.
- $\mathbf{P}(\xi \in D \mid \eta) = \mathbf{P}(\xi \in D)$ *for any Borel set* $D \subset \mathbf{R}^n$.

In particular, if ξ and η are independent random variables, then $\mathbf{E}\xi\eta = \mathbf{E}\xi\mathbf{E}\eta$.

Example 1.52 It can happen that $\mathbf{E}\xi\eta = \mathbf{E}\xi\mathbf{E}\eta$, but ξ and η are not independent. For instance, take $\eta, \zeta \sim N(0, 1)$ such that ζ and η are independent, and take $\xi = \zeta\eta$, then $\mathbf{E}\xi\eta = \mathbf{E}\xi\mathbf{E}\eta = 0$ but ξ depends on η.

Definition 1.53 *Random events* $A_1, A_2, \dots, A_n$ *are said to be mutually independent if* $\mathbf{P}(A_{i_1}A_{i_2}\cdots A_{i_m}) = \mathbf{P}(A_{i_1})\mathbf{P}(A_{i_2})\cdots\mathbf{P}(A_{i_m})$ *for every* $m = 2, \dots, n$ *and for every subset of indices* $i_1, \dots, i_m$.

Definition 1.54 *Random vectors* $\xi_1, \xi_2, \dots, \xi_n$ *are said to be mutually independent if the random events* $\{\xi_1 \in D_1\}, \dots, \{\xi_n \in D_n\}$ *are mutually independent for any Borel sets* $D_1, \dots, D_n$ *(or any open sets).*

1.8 Probability distributions

In most common undergraduate courses in probability theory, random variables are studied via their probability distributions. It is the classical approach. (However, we also need the axiomatic approach described above.)

Distribution theory is to characterize and describe random variables and vectors in terms of the distribution of their possible values and generated measures.

A *probability distribution* on $\mathbf{R}^n$ is a probability measure on the σ-algebra of Borel subsets; i.e., it assigns a probability to every Borel set, so that the probability axioms are satisfied. In fact, this measure is uniquely defined by its values for sets $\{x \in \mathbf{R}^n : a_i < x_i \le b_i\}$ that generate the Borel σ-algebra; if $n = 1$, then the Borel σ-algebra is generated by intervals. Every random vector gives rise to a probability distribution, and this distribution contains most of the important information about the random vector. On the other hand, for any probability distribution, one can find a random vector with that distribution.

Let us consider the simplest case of one-dimensional probability distribution, when the distribution is defined on $\mathbf{R}$. In this case, the distribution assigns to every interval of the real numbers a probability. If ξ is a random variable, the corresponding probability distribution assigns to the interval $(a, b]$ the probability $\mathbf{P}(a < \xi \le b)$, i.e., the probability that the variable ξ will take a value in the interval $(a, b]$; the probability distribution of a random variable completely describes its probabilistic properties. The distributions of any real-valued random variable ξ is uniquely defined by the *cumulative distribution function* (c.d.f.) $F(x) = \mathbf{P}(\xi \le x)$, where $-\infty < x < +\infty$. We have that $\mathbf{P}(a < \xi \le b) = F(b) - F(a)$ and $\mathbf{P}(\xi = x) = F(x) - F(x - 0)$ for any real x. This function F is non-decreasing, $F(x + 0) = F(x)$ for all x, $F(-\infty) = 0$, $F(+\infty) = 1$. On the other hand, any function F with these properties uniquely defines a probability distribution.

The support of a distribution is the smallest closed set whose complement has probability zero.

A distribution is called *discrete* if it belongs to a random variable ξ which can only attain values from a certain finite or countable set. This random variable ξ is also said to be discrete. In this case, the distribution can be given in the form of a complete list of the possible realizations, and a specialization of the probability of each. This is called the *point density function* or *probability mass function*. In fact, ξ is discrete if and only if its c.d.f. $F(x)$ is a piecewise constant step function.

A distribution is called *continuous* if its cumulative distribution function is continuous, which means that it belongs to a random variable ξ for which $\mathbf{P}(\xi = x) = 0$ for all $x \in \mathbf{R}$. The corresponding random variable is also said to be continuous. Note that if ξ is a discrete random variable and η is a continuous random variable then $\xi + \eta$ is a random variable which is in general neither discrete nor continuous.

A distribution and the corresponding random variable is called *absolutely continuous* if $F(x)$ is absolutely continuous, i.e., $F(x) = \int_{-\infty}^{x} p(y)dy$ for some integrable function p. The derivative $p(x) = dF(x)/dx$ is said to be the *probability density function* (p.d.f.). In this case

$$\mathbf{P}(a < \xi < b) = \mathbf{P}(a \le \xi < b) = \mathbf{P}(a < \xi \le b) = \mathbf{P}(a \le \xi \le b)$$
$$= \int_a^b p(x)dx = F(b) - F(a)$$

for any $a \le b$, and the distribution of ξ can be described by its p.d.f. Discrete distributions do not admit such a density, which is not too surprising. There are continuous distributions that are not absolutely continuous, i.e. which do not admit a density (an example is a distribution supported on the Cantor set; it is a non-countable set with zero measure).

Note that identity of two probability distributions does not mean identity of the random variables for which they belong. For example, ξ and $-\xi$ have the same distribution if ξ is *standard normal*.

The key concept of the theory of random variables and their distributions is the *expectation*, or *mean*. If ξ is discrete, then the expectation $\mathbf{E}\xi$ is

$$\mathbf{E}\xi = \sum_k x_k \mathbf{P}(\xi = x_k) = \sum_k x_k [F(x_k) - F(x_k - 0)],$$

provided that the series is absolutely convergent. Here $\{x_k\}$ is the set of all possible values of ξ. Since ξ is discrete, this set is at least countable. If this set is finite, then the expectation $\mathbf{E}\xi$ always exists; if this set is infinite and the series is not converging absolutely, we say that the expectation of ξ does not exist. If ξ has p.d.f. $p(x)$, then

$$\mathbf{E}\xi = \int_{-\infty}^{+\infty} xp(x)dx$$

provided that

$$\mathbf{E}|\xi| = \int_{-\infty}^{+\infty} |x|\,p(x)dx < +\infty.$$

If $\int_{-\infty}^{+\infty} |x|p(x)dx = +\infty$, then we say that $\mathbf{E}|\xi| = +\infty$ and that the expectation of ξ does not exist.

In general, the expectation $\mathbf{E}\xi$ can be found as

$$\mathbf{E}\xi = \int_{-\infty}^{+\infty} x dF(x),$$

provided that the integral exists; and the integral exists if and only if

$$\mathbf{E}|\xi| = \int_{-\infty}^{+\infty} |x|dF(x) < +\infty.$$

In this case, the random variable ξ is said to have expectation, or to be integrable. In fact, this definition is equivalent to the previous definition of $\mathbf{E}\xi$ as $\mathbf{E}\xi = \int_{\Omega} \xi(\omega)\mathbf{P}(d\omega)$, where $(\Omega, \mathcal{F}, \mathbf{P})$ is the probability space such that $\xi : \Omega \to \mathbf{R}$ is an $\mathcal{F}$-measurable mapping. Here $\Omega = \{\omega\}$ is the set of elementary events, $\mathcal{F}$ is a σ-algebra of subsets of Ω, $\mathbf{P} : \mathcal{F} \to \mathbf{R}$ is a probability measure.

In particular, for any measurable function $f : \mathbf{R} \to \mathbf{R}$, $\eta = f(\xi)$ is also a random variable, and the expectation $\mathbf{E}\eta = \mathbf{E}f(\xi)$ can be found as

$$\mathbf{E}f(\xi) = \int_{-\infty}^{+\infty} f(x)dF(x),$$

provided that the random variable $f(\xi)$ is integrable. (For example, it holds if f is bounded, or f is continuous and the support of the distribution of ξ is bounded.) If ξ has p.d.f. $p(x)$, then

$$\mathbf{E}f(\xi) = \int_{-\infty}^{+\infty} f(x)p(x)dx.$$

In many problems we are interested in certain characteristics (instead of the complete characterization) of a random variable, for example its mean, variance, median, moments, quantiles, and some others.

The *kth moment* of a random variable is defined as $\mathbf{E}\xi^k$. The *kth absolute moment* of a random variable is defined as $\mathbf{E}|\xi|^k$. If $\mathbf{E}|\xi|^k < +\infty$, then ξ is said to have kth moment. Note that $(\mathbf{E}|\xi|^k)^{1/k} \le (\mathbf{E}|\xi|^m)^{1/m}$ if $0 \le k \le m$. It follows that if $\mathbf{E}|\xi|^m < +\infty$ then $\mathbf{E}|\xi|^k < +\infty$. But it is possible that $\mathbf{E}|\xi|^k < +\infty$ and $\mathbf{E}|\xi|^m = +\infty$ (for example, $\mathbf{E}|\xi| < +\infty$ and $\mathbf{E}\xi^2 = +\infty$).

If the distribution of a random variable is known, then the distributions of all linear transformations of this variable can also be derived. Suppose that a real-valued random variable ξ has c.d.f. F. Let a, b be constants with $b > 0$. Then the linear transformation $\eta = a + b\xi$ has c.d.f. F_η given by $F_\eta(x) = F[(x - a)/(b)]$. Clearly, the distribution of ξ and the pair (a, b) defines the distribution of η. In particular, $\mathbf{E}\eta = a + b\mathbf{E}\xi$, $\text{Var}\,\eta = b^2\text{Var}\,\xi$. If ξ has density function p, then η has density function $p_\eta(x) = b^{-1}p[(x - a)/(b)]$. It is convenient to describe this family via the distribution of ξ with $\mathbf{E}\xi = 0$, $\text{Var}\,\xi = 1$, or *standardized distribution*.

Vector case

Let us consider now the case of n-dimensional probability distribution, i.e. when the distribution is defined on $\mathbf{R}^n$. In this case, the distribution is a probability measure that assigns to every measurable (Borel) set of $\mathbf{R}^n$ a probability. If $\xi = (\xi_1, \ldots, \xi_n)$ is a random vector, the corresponding probability distribution assigns to the rectangle $\prod_{k=1}^n (a_i, b_i]$ the probability $\mathbf{P}(a_k < \xi_k \leq b_k,\ k = 1, \ldots, n)$. This defines the distribution uniquely for the Borel σ-algebra. Hence the distribution of any random vector $\xi = (\xi_1, \ldots, \xi_n)$ is uniquely defined by the cumulative distribution function $F(x_1, \ldots, x_n) = \mathbf{P}(\xi_1 \leq x_1, \ldots, \xi_n \leq x_n)$, where $-\infty < x_i < +\infty$. We call this function also the *joint cumulative distribution function* of random variables $\xi_1, \ldots, \xi_n$. The probability distribution of a random vector completely describes its probabilistic properties, including mutual dependence of components. The components ξ_k are independent if and only if $F(x_1, \ldots, x_n) = \prod_{k=1}^n F_k(x_k)$, where $F_k(x) = \mathbf{P}(\xi_k \leq x)$. In general, the probability distribution of a random vector is not uniquely defined by the set of probability distributions of its components.

Probability distributions on infinite dimensional spaces are commonly used in the theory of stochastic processes. For example, the *Wiener process* $w(t)_{t \in [0,T]}$ studied in Chapter 4 is a random (infinity-dimensional) vector with values at the space $C(0, T)$ of continuous functions $f : [0, T] \to \mathbf{R}$. It generates a probability distribution on this space (the so-called *Wiener measure*).

Some special distributions

The *binomial coefficient* is defined as $C_n^k = n!/k!(n - k)!$; here $m!$ denotes the factorial of m.

Some useful special distributions are listed below.

Discrete uniform distribution

The discrete uniform distribution is a distribution where all elements of a finite set are equally likely. This is supposed to be the distribution of a balanced coin, an unbiased die, or a casino roulette.

Binomial distribution

Let $n > 0$ be an integer, and let $p \in [0, 1]$. A random variable ξ such that

$$\mathbf{P}(\xi = k) = C_n^k p^k (1-p)^{n-k}, \quad k = 1, \ldots, n$$

is said to have the binomial distribution. The number of successes in n repeating *Bernoulli trials* has this distribution. For this distribution, $\mathbf{E}\xi = np$ and $\operatorname{Var} \xi = np(1-p)$.

Continuous uniform distribution

A random variable ξ with density

$$p(x) = \begin{cases} (b-a)^{-1}, & x \in (a, b) \\ 0, & \text{otherwise}, \end{cases}$$

is said to have the uniform distribution on $[a, b]$, $a < b$. For this distribution, all points in a finite interval are equally likely, and $\mathbf{E}\xi = (b-a)/2$, $\operatorname{Var} \xi = (b-a)^2/12$. In many program languages, built-in random number generators produce the uniform distribution on $[0, 1]$ (in MATLAB, it is the *rand* command).

Normal or Gaussian distribution

A random variable ξ with density

$$p(x) = \frac{1}{\sigma\sqrt{2\pi}} \exp\left\{-\frac{(x-a)^2}{2\sigma^2}\right\}, \quad x \in \mathbf{R},$$

is said to be a Gaussian random variable. It is said also to have the *normal* or *Gaussian distribution*. It can be written as $\xi \sim N(a, \sigma^2)$. For this distribution, $\mathbf{E}\xi = a$ and $\operatorname{Var} \xi = \sigma^2$. If $a = 0$ and $\sigma = 1$, then the distribution is said to be the standard normal distribution. This is an extremely important probability distribution which has applications in statistics, probability theory, physics, and engineering. By the *central limit theorem*, it is the limit distribution for the mixing of a large number of independent random variables.

Log-normal distribution

Let ξ be a Gaussian random variable, i.e., $\xi \sim N(a, \sigma^2)$, where $a \in \mathbf{R}$, $\sigma > 0$. A random variable $\eta = e^\xi$ is said to have log-normal distribution (i.e., $\ln \eta$ is normal). The random variable η has density function

$$p(x) = \begin{cases} \dfrac{1}{x\sigma\sqrt{2\pi}} \exp\left\{-\dfrac{(\ln x - a)^2}{2\sigma^2}\right\}, & x > 0 \\ 0, & x \le 0 \end{cases} \tag{1.1}$$

For this distribution, $\mathbf{E}\eta = \exp\{a + \sigma^2/2\}$, $\mathbf{E}\eta^2 = \exp\{2a + 2\sigma^2\}$ and $\mathrm{Var}\,\eta = \exp\{2a + 2\sigma^2\} - \exp\{2a + \sigma^2\}$.

1.9 Problems

Solve Problems 1.29 and 1.40.

Problem 1.55 Describe the difference between an event and an elementary event in a probability space. Is a probability of occurrence assigned to an elementary event?

Problem 1.56 Find an example of a probability space with $\Omega = \{a, b, c, d\}$, an example of a (non-constant) random variable ξ, and find $\mathbf{E}\xi$, $\mathbf{E}\xi^2$, and $\mathrm{Var}\,\xi$. Find its cumulative distribution function (c.d.f.) $F(x) = \mathbf{P}(\xi \le x)$.

Problem 1.57 Let $\Omega = \{\omega_1, \omega_2, \omega_3\}$, $\mathcal{F} = 2^\Omega$. Let $\xi(\omega_i) = (i-2)^2$. Find the σ-algebra $\sigma(\xi)$ generated by ξ.

Problem 1.58 Let $(\Omega, \mathcal{F}, \mathbf{P})$ be a probability space such that $\Omega = \{\omega_1, \omega_2, \omega_3\}$, $\mathcal{F} = 2^\Omega$,

$$\mathbf{P}(\{\omega_1\}) = 1/4, \quad \mathbf{P}(\{\omega_2\}) = 1/4, \quad \mathbf{P}(\{\omega_3\}) = 1/2.$$

Let $\xi(\omega_1) = 0.1$, $\xi(\omega_2) = 0$, $\xi(\omega_3) = -0.1$. Let $\eta(\omega_1) = \eta(\omega_3) = 0.1$, $\eta(\omega_2) = -0.1$. Find $\mathbf{E}\{\xi\,|\,\eta\}$.

Challenging problems

Problem 1.59 Is there an example of a probability distribution on $\mathbf{R}$ such that its support coincides with the set of all rational numbers? If yes, give an example; if no, prove it.

Problem 1.60 Let $\mathbf{Q}^2$ be the set of all pairs $(x, y) \in \mathbf{R}^2$ such that x and y are rational numbers. We consider a random direct line L in $\mathbf{R}^2$ such that $(0, 0) \in L$ with probability 1, and that the angle between L and the vector $(1, 0)$ has the uniform distribution on $[0, \pi)$. Find the probability that the set $L \cap \mathbf{Q}^2$ is finite.

2 Basics of stochastic processes

In this chapter, some basic facts and definitions from the theory of stochastic (random) processes are given, including filtrations, martingales, Markov times, and Markov processes.

2.1 Definitions of stochastic processes

Sometimes it is necessary to consider random variables or vectors that depend on time.

Definition 2.1 *A sequence of random variables ξ_t, $t = 0, 1, 2, \dots$, is said to be a discrete time stochastic (or random) process.*

Definition 2.2 *Let $T \in [0, +\infty]$ be given. A mapping $\xi : [0, T] \times \Omega \to \mathbf{R}$ is said to be a continuous time stochastic (random) process if $\xi(t, \omega)$ is a random variable for a.e. (almost every) t.*

A random process has two independent variables (t and ω). It can be written as $\xi_t(\omega)$, $\xi(t, \omega)$, or just ξ_t, $\xi(t)$.

Example 2.3 Let η be a random variable. Then $\xi_1(t) \triangleq t + \eta$ and $\xi_2(t) = \sin(\eta t)$ are random processes.

Example 2.4 Consider an ordinary differential equation

$$\frac{dy}{dt}(t) = f(y(t), t),$$
$$y(0) = a(\omega),$$

when $a(\omega)$ is a random variable. Then the solution $y(t) = y(t, \omega)$ is a continuous time random process (provided that this solution exists and it is well defined).

It can be seen that the randomness is presented only in initial time $t = 0$ for the process from the last example, and the evolution of this process is uniquely defined by its initial data. The following definitions give examples that are different.

Definition 2.5 *Let ξ_t, $t = 0, 1, 2, \dots$, be a discrete time random process such that ξ_t are mutually independent and have the same distribution, and $\mathbf{E}\xi_t \equiv 0$. Then the process ξ_t is said to be a discrete time white noise.*

Definition 2.6 *Let ξ_t be a discrete time white noise, and let $\eta_t \stackrel{\Delta}{=} \xi_0 + \xi_1 + \cdots + \xi_t$, $t = 0, 1, 2, \dots$. Then the process η_t is said to be a random walk.*

The theory of stochastic processes studies their *pathwise properties* (or properties of trajectories $\xi(t, \omega)$ for given ω, as well as the evolution of the probability distributions.

Definition 2.7 *A continuous time process $\xi(t) = \xi(t, \omega)$ is said to be continuous (or pathwise continuous), if trajectories $\xi(t, \omega)$ are continuous in t a.s. (i.e., with probability 1, or for a.e. ω).*

It can happen that a continuous time process is not continuous (for instance, a process with jumps).

2.2 Filtrations, independent processes and martingales

In this section, we shall assume that either $t \in [0, +\infty)$ or $t = 0, 1, 2, \dots$.

Filtrations

In addition to evolving random variables, we shall use evolving σ-algebras.

Definition 2.8 *A set of σ-algebras $\{\mathcal{F}_t\}$ is called a filtration if $\mathcal{F}_s \subseteq \mathcal{F}_t$ for $s < t$.*

Definition 2.9 *Let $\xi(t)$ be a random process, and let $\mathcal{F}_t$ be a filtration. We say that the process $\xi(\cdot)$ is adapted to the filtration $\mathcal{F}_\cdot$, if any random variable $\xi(t)$ is measurable with respect to $\mathcal{F}_t$ (i.e., $\{\xi(t) \in B\} \in \mathcal{F}$, where $B \subset \mathbf{R}$ is any open interval).*

Definition 2.10 *Let $\xi(t)$ be a random process. The filtration $\mathcal{F}_t$ generated by $\xi(t)$ is defined as the minimal filtration such that $\xi(t)$ is adapted to it.*

Example 2.11 Let $\Omega = \{\omega_1, \omega_2, \omega_3\}$, $\mathcal{F} = 2^\Omega$. Consider a discrete time random process ξ_t, $t = 0, 1, \dots$ such that

$$\xi_0(\omega) = \begin{cases} 0 & \omega = \omega_2 \\ 1 & \omega = \omega_1 \text{ or } \omega = \omega_3 \end{cases}$$

$$\xi_1(\omega) = \begin{cases} 1.1 & \omega = \omega_2 \\ 1 & \omega = \omega_1 \text{ or } \omega = \omega_3 \end{cases} \qquad \xi_2(\omega) = \begin{cases} 7 & \omega = \omega_1 \\ 1.1 & \omega = \omega_2 \\ 0.5 & \omega = \omega_3 \end{cases}$$

Let us find the filtration $\mathcal{F}_t$ generated by ξ_t for $t = 0, 1, 2$.

We have that ξ_i generates the σ-algebra $\mathcal{F}_{\xi_i} = \{\Omega, \emptyset, \{\omega_2\}, \{\omega_1, \omega_3\}\}$, $i = 0, 1$, and ξ_2 generates the σ-algebra $\mathcal{F}_{\xi_2} = 2^\Omega$ of all subsets of Ω. We have that $\mathcal{F}_{\xi_0} = \mathcal{F}_{\xi_1} \subseteq \mathcal{F}_{\xi_2}$, so we can conclude that $\mathcal{F}_{\xi_t}$ is the filtration generated by ξ_t: the filtration generated by ξ_t cannot consist of smaller σ-algebras, and $\mathcal{F}_{\xi_0} \subseteq \mathcal{F}_{\xi_1} \subseteq \mathcal{F}_{\xi_2}$.

Example 2.12 Let $\Omega = \{\omega_1, \omega_2, \omega_3\}$, $\mathcal{F} = 2^\Omega$. Consider a discrete time random process ξ_t, $t = 0, 1, \ldots$ such that

$$\xi_0(\omega) = \begin{cases} 0 & \omega = \omega_2 \\ 1 & \omega = \omega_1 \text{or} \omega = \omega_3 \end{cases} \qquad \xi_1(\omega) = \begin{cases} -7 & \omega = \omega_1 \\ 0.3333 & \omega = \omega_2 \\ 0.5 & \omega = \omega_3 \end{cases}$$

$$\xi_2(\omega) = \xi_0(\omega).$$

Let us find again the filtration $\mathcal{F}_t$ generated by the process ξ_t, $t = 0, 1, 2$.

Let $\mathcal{F}_{\xi_t}$ denote the σ-algebra generated by the random variable ξ_t for any given time t. We have that ξ_i generates the σ-algebra $\mathcal{F}_{\xi_i} = \{\Omega, \emptyset, \{\omega_2\}, \{\omega_1, \omega_3\}\}$, $i = 0, 2$, and ξ_1 generates the σ-algebra $\mathcal{F}_{\xi_1} = 2^\Omega$ of all subsets of Ω. Therefore, the filtration generated by the process ξ_t is $\mathcal{F}_0 = \mathcal{F}_{\xi_0}$, $\mathcal{F}_1 = \mathcal{F}_{\xi_1} = 2^\Omega$, $\mathcal{F}_2 = \mathcal{F}_1$. (Note that $\mathcal{F}_{\xi_1} \nsubseteq \mathcal{F}_{\xi_2}$, i.e., the sequence $\{\mathcal{F}_{\xi_t}\}_{t \geq 0}$ is not 'non-decreasing'), therefore $\{\mathcal{F}_{\xi_t}\}_{t \geq 0}$ is not a filtration; to make the sequence non-decreasing, we must replace $\mathcal{F}_{\xi_2}$ by $\mathcal{F}_{\xi_1} = 2^\Omega$).

Independent processes

Definition 2.13 *Random processes $\xi(\cdot)$ and $\eta(\cdot)$ are said to be independent iff the events $\{(\xi(t_1), \ldots, \xi(t_n)) \in A\}$ and $\{(\eta(\tau_1), \ldots, \eta(\tau_m)) \in B\}$ are independent for all m, n, all times $(t_1, \ldots, t_n)$ and $(\tau_1, \ldots, \tau_m)$, and all sets $A \subset \mathbf{R}^n$ and $B \subset \mathbf{R}^m$.*

In fact, processes are independent iff all events from the filtrations generated by them are mutually independent.

Martingales

Definition 2.14 *Let $\xi(t)$ be a process such that $\mathbf{E}|\xi(t)|^2 < +\infty$ for all t, and let $\mathcal{F}_t$ be a filtration. We say that $\xi(t)$ is a martingale with respect to $\mathcal{F}_t$ if*

$$\mathbf{E}\{\xi(t)|\mathcal{F}_s\} = \xi(s) \quad a.s. \quad \forall s, t \colon s < t.$$

Note that we require that $\mathbf{E}|\xi(t)|^2 < +\infty$ because, for simplicity, we have defined the conditional expectation only for this case. In the literature, the martingales are often defined under the condition $\mathbf{E}|\xi(t)| < +\infty$, which is less restrictive.

Sometimes the term 'martingale' is used without mentioning the filtration.

Definition 2.15 *Let $\xi(t)$ be a process, and let $\mathcal{F}_t^\xi$ be the filtration generated by this process. We say that $\xi(t)$ is a martingale if $\xi(t)$ is a martingale with respect to the filtration $\mathcal{F}_t^\xi$.*

Problem 2.16 Prove that any discrete time random walk is a martingale.

Problem 2.17 Let ζ be a random variable such that $\mathbf{E}|\zeta|^2 < +\infty$, and let $\mathcal{F}_t$ be a filtration. Prove that $\xi(t) \triangleq \mathbf{E}\{\zeta|\mathcal{F}_t\}$ is a martingale with respect to $\mathcal{F}_t$.

The following definitions will also be useful.

Definition 2.18 *Let $\xi(t)$ be a process such that $\mathbf{E}|\xi(t)|^2 < +\infty$ for all t, and let $\mathcal{F}_t$ be a filtration. We say that $\xi(t)$ is a submartingale with respect to $\mathcal{F}_t$ if*

$$\xi(s) \le \mathbf{E}\{\xi(t)|\mathcal{F}_s\} \quad a.s. \quad \forall s,t\colon s < t.$$

Definition 2.19 *Let $\xi(t)$ be a process such that $\mathbf{E}|\xi(t)|^2 < +\infty$ for all t, and let $\mathcal{F}_t$ be a filtration. We say that $\xi(t)$ is a supermartingale with respect to $\mathcal{F}_t$ if*

$$\xi(s) \ge \mathbf{E}\{\xi(t)|\mathcal{F}_s\} \quad a.s. \quad \forall s,t\colon s < t.$$

The term *sub(super)martingale* can also be used without mentioning the filtration, meaning the filtration generated by the process itself.

2.3 Markov times

To discuss American options, we need some additional definitions for random times.

Let $\mathcal{F}_t$ be a filtration.

Definition 2.20 *Markov time with respect to $\mathcal{F}_t$ is any random time τ such that $\{\tau \le t\} \in \mathcal{F}_t$ for all t.*

The definition means that, for a Markov time τ, we can say at time t if $\tau \le t$ or not if we know all events from $\mathcal{F}_t$.

In particular, if $\mathcal{F}_t$ is the filtration generated by a random process $\xi(t)$ and τ is a Markov time with respect to $\mathcal{F}_t$, then we can say at time t if $\tau \le t$ or not if we know all values $\{\xi(s), s \le t\}$.

Corollary 2.21 *τ is a Markov time iff the process $\mathbb{I}_{\{t\le\tau\}}$ is $\mathcal{F}_t$-adapted.*

For instance, $\tau = \min\{t \ge 0\colon \xi(t) = 2\}$ is Markov time with respect to the filtration $\mathcal{F}_t$ generated by a process $\xi(t)$, but τ such that $\xi(\tau) = \max_{s\in[0,1]} \xi(s)$ is not a Markov time. In particular, the mathematical concept of Markov time explains why one cannot catch the best time to sell stocks (when the price is maximal): this time exists, but we cannot catch it by observing current data. The same is true for the best time of exit from stochastic games like roulette.

Sometimes Markov times are called stopping times.

Definition 2.22 *Let τ be a Markov time, then $\mathcal{F}_\tau \stackrel{\Delta}{=} \{A \in 2^\Omega: A \cap \{t \le \tau\} \in \mathcal{F}_t \, \forall t\}$.*

It can be shown that $\mathcal{F}_\tau$ is a σ-algebra, and that τ and $\xi(\tau)$ are $\mathcal{F}_\tau$-measurable with respect to it, for any $\mathcal{F}_t$-adapted process $\xi(t)$ (in fact, for the case of continuous time process $\xi(t)$, it is true under some additional conditions; it suffices to require that $\xi(t)$ is pathwise continuous).

2.4 Markov processes

Definition 2.23 *Let $\xi(t)$ be a process, and let $\mathcal{F}_t^\xi$ be the filtration generated by $\xi(t)$. We say that $\xi(t)$ is a Markov (Markovian) process if*

$$\mathbf{P}(\xi(t_1) \in D_1, \dots, \xi(t_k) \in D_k \mid \mathcal{F}_s^\xi) = \mathbf{P}(\xi(t_1) \in D_1, \dots, \xi(t_k) \in D_k \mid \xi(s))$$

for any $k > 0$, for any times s and t_m such that $t_m > s$, for any system of open sets $\{D_m\}$, $m = 1, \dots, k$.

This property is said to be the Markov property.

The Markov property means that if we want to estimate the distribution of $\xi(t)|_{t>s}$ using information of the past values $\xi(r)|_{r \le s}$, it suffices to use the last observable value $\xi(s)$ only. Using the values for $r \in [0, s)$ does not give any additional benefits. This property (if it holds) helps to solve many problems.

The following proposition will be useful.

Proposition 2.24 *Under the assumptions and notations of Definition 2.23, $\mathbf{E}\{F(\xi(t_1), \xi(t_2), \dots, \xi(t_k)) \mid \mathcal{F}_s^\xi\} = \mathbf{E}\{F(\xi(t_1), \xi(t_2), \dots, \xi(t_k)) \mid \xi(s)\}$ for all measurable deterministic functions F such that the corresponding random variables are integrable.*

Problem 2.25 Prove that a discrete time random walk is a Markov process.

Vector processes

Let $\xi(t) = (\xi_1(t), \dots, \xi_n(t))$ be a vector process such that all its components are random processes. Then ξ is said to be an n-dimensional (vector) random process. All definitions given above can be extended for these vector processes.

Sometimes, we can convert a process that is not a Markov process to a Markov process of higher dimension.

Example 2.26 Let η_t be a random walk, $t = 0, 1, 2, \ldots$, and let $\psi_t \triangleq \eta_1 + \cdots + \eta_t$. Then ψ_t is not a Markov process, but the vector process (η_t, ψ_t) is a Markov process.

2.5 Problems

Problem 2.27 Let ζ be a random variable, and let $0 \le a < b \le 1$. Let a continuous time random process $\xi(t)$ be such that

$$\xi(t) = \begin{cases} \zeta, & t \in [a,b] \\ 1, & t \notin [a,b] \end{cases}$$

Find the filtration $\mathcal{F}_t^\xi$ generated by $\xi(t)$ for $t \in [0,1]$ for the cases when $a = 0$, $b = 1/2$, and when $a = 1/4$, $b = 2/3$.

Problem 2.28 Let $\Omega = \{\omega_1, \omega_2, \omega_3\}$, $\mathcal{F} = 2^\Omega$. Consider a discrete time random process ξ_t, $t = 0, 1, \ldots$. Let $\xi_t(\omega_i) = (i-2)^{2+t(t-1)/2}$. Find the filtration $\mathcal{F}_t^\xi$ generated by $\xi(t)$ for $t = 0, 1, 2$.

Problem 2.29 Let ζ be a random variable. Let the random process $\xi(t)$ be such that

$$\xi(t) = \begin{cases} \zeta, & t \in [0,1] \\ 1, & t > 1 \end{cases}$$

Find the filtration $\mathcal{F}_t$ generated by $\xi(t)$ for $t \in [0,2]$.

Problem 2.30 Let $(a,b) \in \mathbf{R}^2$. Let a process ξ_t be such that

$$\mathbf{P}(\xi_t \in \{a,b\}) = 1, \quad \mathbf{P}(\xi_t = a) > 0, \quad \mathbf{P}(\xi_t = b) > 0 \quad (\forall t).$$

Let $\eta_t \triangleq \xi_1 + \cdots + \xi_t$. Is it possible to find a measure $\mathbf{P}$ such that η_t is a martingale? If yes, give an example. Consider the cases: (i) $a = -1,\ b = 1$; (ii) $a = 1,\ b = 1/2$; (iii) $a = b = -1$.

Hint: in Problems 2.30 and 2.31, look for a measure in the class of measures such that ξ_t are independent.

Problem 2.31 Let $(a,b,c) \in \mathbf{R}^3$. Let a process ξ_t be such that

$$\mathbf{P}(\xi_t \in \{a,b,c\}) = 1, \quad \mathbf{P}(\xi_t = a) > 0, \quad \mathbf{P}(\xi_t = b) > 0, \quad \mathbf{P}(\xi_t = c) > 0 \quad (\forall t).$$

Let $\eta_t \triangleq \xi_1 + \cdots + \xi_t$. Is it possible to find a measure $\mathbf{P}$ such that η_t is a martingale? If yes, give an example. Consider the following cases: (i) $a = -1,\ b = 2,\ c = 3$; (ii) $a = 1,\ b = 2,\ c = 3$; (iii) $a = b = -1,\ c = 2$.

Problem 2.32 Find when the desired measure **P** is unique in Problems 2.30 and 2.31 (in the class of measures such that ξ_t are independent).

Problem 2.33 Let τ be a Markov time with respect to a filtration. (i) Is 2τ a Markov time? (ii) Is $\tau/2$ a Markov time?

3 Discrete time market models

In this chapter, we study discrete time mathematical models of markets. These models are relatively simple and straightforward. However, they still allow us to introduce all core definitions of mathematical finance such as *self-financing strategies, replicating, arbitrage, risk-neutral measures, market completeness, and option price*. Besides, these models have fundamental significance, and their theory is not completed yet.

3.1 Introduction: basic problems for market models

Market models are being created with the goal of explaining the internal logic of market transactions and the laws of price movement. In addition, market models target the following key problems:

- *Portfolio selection problem: To find a strategy of buying and selling stocks.*
- *Pricing problem: To find a 'fair' price for derivatives (i.e., options, futures, etc.).*

There is an auxiliary problem (which is the main problem for *financial econometrics*):

- *To estimate the evolution law of the probability distributions from market statistics.*

Portfolio selection problem

A generic *optimal investment (or optimal portfolio selection) problem* is

$$\text{Maximize} \quad \mathbf{E}U(X(T)) \quad \text{over a class of investment strategies.}$$

Here T is the terminal time, $X(T)$ is the wealth of an investor at time T, and $U(\cdot)$ is a given *utility function* that describes risk preferences. Typically, the stock prices, the wealth, and the strategies, are supposed to be random processes; this is why there is maximization for the expectation $\mathbf{E}U$.

The most common utilities are log and power functions, i.e., $U(x) = \ln x$ and $U(x) = \delta^{-1}x^{\delta}$, where $\delta < 1$ and $d \neq 0$. Another important example of utility is $U(x) = kx - \mu x^2$, where $k > 0$ and $\mu > 0$ are some constants.

This generic problem allows many modifications, including:

- optimal strategies of consumption and dividends;
- optimal hedging of (non-replicable) claims;
- problem with constraints on the wealth;
- problems with $T = +\infty$.

The theory of optimal portfolio selection can be considered as a special part of *optimal control theory*, or, more precisely, of *optimal stochastic control theory*. Optimal investment problems are intensively studied in the literature. However, they are not discussed here in detail, because this course is focused on the problem of pricing (the only exception is Section 3.12).

3.2 Discrete time model with free borrowing

We introduce a model of a financial market consisting of the risky stock with price S_t, $t = 0, 1, 2, \ldots$, where t are times (for example, days, weeks, months, etc.). The initial price $S_0 > 0$ is a given non-random value.

Let us assume first that the rate for money borrowing and lending is zero.

Let us describe investment operations, or portfolio strategies.

Let $X_0 > 0$ be the initial wealth at time $t = 0$, and let X_t be the wealth at time $t \geq 0$.

We assume that the wealth X_t at time $t \geq 0$ is

$$X_t = \beta_t + \gamma_t S_t, \quad t = 0, 1, 2, \ldots, \tag{3.1}$$

where β_t is the quantity of cash on a bank account, and γ_t is the quantity of the stock portfolio. The pair (β_t, γ_t) describes the state of the portfolio at time t. We call a sequence of these pairs strategy, or portfolio strategy.

Note that we allow negative β_t and γ_t, meaning borrowing or short positions.

Some constraints will be imposed on strategies.

Definition 3.1 *A sequence* $\{(\beta_t, \gamma_t)\}$ *is said to be an admissible strategy if there exist measurable functions* $F_t : \mathbf{R}^{t+1} \to \mathbf{R}^2$ *such that*

$$(\beta_t, \gamma_t) = F_t(S_0, S_1, \ldots, S_t).$$

It follows from this definition that the process (β_t, γ_t) is adapted to the filtration generated by S_t, and that (β_t, γ_t) does not use information about the 'future', or about S_{t+m} for $m > 0$.

Definition 3.2 *We say that the strategy is self-financing if*

$$X_{t+1} - X_t = \gamma_t (S_{t+1} - S_t), \quad t = 0, 1, \ldots . \tag{3.2}$$

It follows from (3.2) that

$$X_t = X_0 + \sum_{m=0}^{t-1} \gamma_m (S_{m+1} - S_m). \tag{3.3}$$

Here $X_0 > 0$ is the initial wealth at time $t = 0$.

For example, for the trivial risk-free strategy, when $\gamma_t \equiv 0$, the corresponding total wealth is $X_t \equiv X_0$.

Note that these definitions present a simplification of the real market situation, because transaction costs, bid and ask gap, possible taxes and dividends, interest rate for borrowing, etc., are not taken into account.

3.3 A discrete time bond–stock market model

A more realistic model of the market with non-zero interest rate for borrowing can be described via the following bond–stock model.

We introduce a model of a market, consisting of the risk-free bond or bank account with price B_t and the risky stock with the price S_t, $t = 0, 1, 2, \ldots$. The initial prices $S_0 > 0$ and $B_0 > 0$ are given non-random variables.

Set

$$\rho_t \stackrel{\Delta}{=} \frac{B_t}{B_{t-1}}, \qquad \xi_t \stackrel{\Delta}{=} \frac{S_t}{\rho_t S_{t-1}} - 1.$$

In other words,

$$\begin{cases} S_t = \rho_t S_{t-1}(1 + \xi_t), \\ B_t = \rho_t B_{t-1}, \end{cases} \qquad t = 1, 2, \ldots . \tag{3.4}$$

We assume that

$$|\xi_t| < 1, \quad \rho_t \geq 1 \quad \forall t. \tag{3.5}$$

Note that these conditions are technical. In particular, they ensure that $S_t > 0$. It is not too restrictive. For instance, if the change in the stock prices is no more than 5% per time period, then $|\xi_t| < 0.05$.

In the case of daily transactions,

$$\rho_t = 1 + \text{interest rate}/365.$$

Remark 3.3 The case of $\rho_t \equiv 1$ corresponds to the market model with free borrowing.

Let $X_0 > 0$ be the initial wealth at time $t = 0$, and let X_t be the wealth at time $t \geq 0$. We assume that the wealth X_t at time $t = 0, 1, 2, \dots$ is

$$X_t = \beta_t B_t + \gamma_t S_t, \tag{3.6}$$

where β_t is the quantity of the bond portfolio, and γ_t is the quantity of the stock portfolio. The pair (β_t, γ_t) describes the state of the bond–stocks securities portfolio at time t. We call sequences of these pairs strategies.

Some constraints will be imposed on strategies.

Note that we allow negative β_t and γ_t, meaning borrowing or short positions.

Definition 3.4 *A sequence* $\{(\beta_t, \gamma_t)\}$ *is said to be an admissible strategy if there exist measurable functions* $F_t : \mathbf{R}^{2t+2} \to \mathbf{R}^2$ *such that*

$$(\beta_t, \gamma_t) = F_t(S_0, B_0, S_1, B_1, \dots, S_t, B_t).$$

It follows from this definition that the process (β_t, γ_t) is adapted to the filtration generated by (S_t, B_t), and that (β_t, γ_t) does not use information about the 'future', or about (S_{t+m}, B_{t+m}) for $m > 0$.

The main constraint in choosing a strategy is the so-called condition of self-financing.

Definition 3.5 *A strategy* $\{(\beta_t, \gamma_t)\}$ *is said to be self-financing, if*

$$X_{t+1} - X_t = \beta_t (B_{t+1} - B_t) + \gamma_t (S_{t+1} - S_t)\,. \tag{3.7}$$

Remark 3.6 In the literature, a definition of admissible strategies may include requirements that the risk is bounded. An example of this requirement is the following: there exists a constant C such that $X_t \geq C$ for all t a.s. For simplicity, we do not require this, because this condition is always satisfied for the special problems discussed below.

Remark 3.7 Similarly, we can consider a multistock market model, when $S_t = \{S_{it}\}$ and $\gamma_t = \{\gamma_{it}\}$ are vectors, and when the wealth is $X_t = \beta_t B_t + \sum_i \gamma_{it} S_{it}$.

Some strategies

Example 3.8 A risk-free ('keep-only bonds') strategy is a strategy when the portfolio contains only the bonds, $\gamma_t \equiv 0$, and the corresponding total wealth is $X_t \equiv \beta_0 B_t \equiv \prod_{m=1}^{t} \rho_m X_0$.

Example 3.9 A *buy-and-hold* strategy is a strategy when $\gamma_t > 0$ does not depend on time. This strategy ensures a gain when stock price is increasing.

Example 3.10 A *short position* is the state of the portfolio when $\gamma_t < 0$. This portfolio ensures a gain when stock price is decreasing.

Example 3.11 A *'doubling strategy'* is sometimes used by an aggressive gambler (for instance in the coin-tossing game). In fact, the stochastic market model is close to the model of gambling. Therefore, it is possible to suggest the analogue of this strategy for the stock market. Let us assume that $S_{t+1} = S_t(1 + \xi_{t+1})$, where $\xi_t = \pm\varepsilon$ is random, with given $\varepsilon > 0$. An analogue of the doubling strategy is as follows: $\gamma_t = S_t^{-1}2^{t+1}$, $t < \tau$, and $\gamma_t = 0$, $t \geq \tau$, where $\tau = \min\{t\colon \xi_t = +\varepsilon\}$.

Problem 3.12 Assume that, with probability 1, there exists a time $\tau = \tau(\omega) \in \{0, 1, 2, \dots\}$ such that $\xi_\tau > 0$. Prove that the doubling strategy ensures with probability 1 positive gain on the unlimited time horizon.

Example 3.13 *'Constantly rebalanced portfolio'* is a strategy such that there is a given constant $C > 0$ such that $\gamma_t S_t / X_t = C$ and $\beta_t B_t / X_t = 1 - C$. In other words, the investor keeps the constant proportion of investment in the bonds and in the stock. This strategy requires selling the stock when its price is going up and buying when it is going down. Therefore, this strategy makes a profit when stock prices oscillate.

Let us describe the resulting wealth for the constantly rebalanced portfolio given C. For simplicity, we assume that $\rho_t \equiv 1$. Then

$$X_{t+1} - X_t = \gamma_t(S_{t+1} - S_t) = \gamma_t S_t \xi_{t+1} = C X_t \xi_{t+1},$$

i.e.,

$$X_{t+1} = X_t(1 + C\xi_{t+1}) = X_{t-1}(1 + C\xi_t)(1 + C\xi_{t+1})$$

$$= \cdots = X_0 \prod_{k=1}^{t+1} (1 + C\xi_k).$$

For instance, let $(S_0, S_1, S_2, S_3, \dots) = (1, 2, 1, 2, 1, \dots)$. It follows that $(\xi_1, \xi_2, \dots) = (1, -\frac{1}{2}, 1, -\frac{1}{2}, 1, -\frac{1}{2}, \dots)$. Let $X_0 = 1$. For the buy-and-hold strategy, the wealth is $(X_1, X_2, X_3, \dots) = (1, 2, 1, 2, 1, \dots)$. In contrast, the constantly rebalanced portfolio with $C = \frac{1}{2}$ gives the wealth $(X_1, X_2, X_3, \dots) = (\frac{3}{2}, \frac{3}{2}\frac{3}{4}, \frac{3}{2}\frac{3}{2}\frac{3}{4}, \dots)$ of an exponential order of growth. (Of course, one cannot be sure that the stock prices will evolve in this specific way.)

Problem 3.14 Consider a discrete time bond–stock market such that $S_0 = 1$, $S_1 = 1.3$, $S_2 = 1.1$. Let the bond prices be $B_0 = 1$, $B_1 = 1.1B_0$, $B_2 = 1.05B_1$.

Let the initial wealth be $X_0 = 1$. Let a self-financing strategy be such that the number of stock shares at the initial time is $\gamma_0 = \frac{1}{2}$. Find $\gamma_1, X_1, X_2, \beta_i, i = 0, 1, 2$ for the constantly rebalanced portfolio.

Solution. We have

$$\gamma_0 S_0 = \frac{X_0}{2} = \frac{1}{2}, \quad C = \tfrac{1}{2}, \quad \beta_0 B_0 = 1 - \tfrac{1}{2} = \tfrac{1}{2}, \qquad \beta_0 = \tfrac{1}{2},$$
$$X_1 = X_0 + 0.3 \cdot \tfrac{1}{2} + 0.1 \cdot \tfrac{1}{2} = 1 + 0.15 + 0.05 = 1.2,$$
$$\gamma_1 S_1 = C \cdot X_1 = \tfrac{1}{2} X_1 = 0.6, \quad \gamma_1 = \frac{0.6}{S_1} = \frac{0.6}{1.3} = 0.462,$$
$$\beta_1 = C \cdot \frac{X_1}{B_1} = \frac{1}{2}\frac{X_1}{1.1} = \frac{0.6}{1.1} = 0.545,$$
$$B_2 = B_0 \cdot 1.1 \cdot 1.05 = 1.155,$$
$$X_2 = 1.2 + (-0.2)\gamma_1 + (1.155 - 1.1)\beta_1$$
$$= 1.2 + (-0.2)0.462 + (0.055)0.545 = 1.138,$$
$$\beta_2 = C\frac{X_2}{B_2} = \frac{X_2}{2B_2} = \frac{1.138}{2 \cdot 1.155} = 0.493. \quad \square$$

3.4 The discounted wealth and stock prices

For the trivial, risk-free, 'keep-only bonds' strategy, the portfolio contains only the bonds, $\gamma_t \equiv 0$, and the corresponding total wealth is $X_t \equiv \beta_0 B_t \equiv \left(\prod_{m=1}^{t} \rho_m\right) X_0$. Some loss is possible for a strategy that deals with risky assets. It is natural to estimate the loss and gain by comparing it with the results for the 'keep-only bonds' strategy.

Definition 3.15 *The process* $\tilde{X}_t \triangleq \left(\prod_{m=1}^{t} \rho_m\right)^{-1} X_t$, $\tilde{X}_0 = X_0$, *is called the discounted wealth (or the normalized wealth).*

Definition 3.16 *The process* $\tilde{S}_t \triangleq \left(\prod_{m=1}^{t} \rho_m\right)^{-1} S_t$, $\tilde{S}_0 = S_0$, *is called the discounted stock price (or the normalized stock price).*

Proposition 3.17 $\tilde{S}_t = \tilde{S}_{t-1}(1 + \xi_t)$.

The proof is straightforward.

Theorem 3.18

$$\tilde{X}_{t+1} - \tilde{X}_t = \gamma_t(\tilde{S}_{t+1} - \tilde{S}_t), \quad t = 0, 1, \ldots \tag{3.8}$$

Proof of Theorem 3.18. Let $\{(\tilde{X}_t, \gamma_t)\}_{t=1}^{n}$ be a sequence such that (3.8) holds. Then it suffices to prove that $X_t \triangleq \left(\prod_{m=1}^{t} \rho_m\right) \tilde{X}_t$ is the wealth corresponding to the self-financing strategy $\{(\beta_t, \gamma_t)\}$, where $\beta_t = (X_t - \gamma_t X_t) B_t^{-1}$.

We have that

$$
\begin{aligned}
X_{t+1} - X_t &= \prod_{m=1}^{t+1} \rho_m \left(\tilde{X}_{t+1} - \tilde{X}_t\right) + (\rho_{t+1} - 1) \prod_{m=1}^{t} \rho_m \tilde{X}_t \\
&= \prod_{m=1}^{t+1} \rho_m \gamma_t \left(\tilde{S}_{t+1} - \tilde{S}_t\right) + (\rho_{t+1} - 1) X_t \\
&= \gamma_t \left(S_{t+1} - \rho_{t+1} S_t\right) + (\rho_{t+1} - 1) X_t \\
&= \gamma_t \left(S_{t+1} - S_t\right) - (\rho_{t+1} - 1) S_t \gamma_t + (\rho_{t+1} - 1) X_t \\
&= \gamma_t \left(S_{t+1} - S_t\right) + (\rho_{t+1} - 1) \left(X_t - S_t \gamma_t\right) \\
&= \gamma_t \left(S_{t+1} - S_t\right) + (\rho_{t+1} - 1) \beta_t B_t \\
&= \gamma_t \left(S_{t+1} - S_t\right) + (B_{t+1} - B_t) \beta_t.
\end{aligned}
$$

This completes the proof. □

Thanks to Theorem 3.18, we can reduce many problems for markets with non-zero interest for borrowing to the simpler case of the market with zero interest rate (i.e., with free borrowing).

For simplicity, one can assume for the first reading that $\rho_t \equiv 1$, $\tilde{X}_t \equiv X_t$, and $\tilde{S}_t \equiv S_t$, everywhere in this chapter. After that, one can read this chapter again taking into account the impact of $\rho_t \neq 1$.

3.5 Risk-neutral measure

Up to this point, we have not needed probability space, and the market model was not a stochastic market model. Now we assume that we are given a standard complete probability space $(\Omega, \mathcal{F}, \mathbf{P})$ (see Chapter 1). The probability measure $\mathbf{P}$ describes the probability distribution of the sequence $\{(\rho_t, \xi_t)\}$. Sometimes we shall address it as the *original probability measure*, or the *prior probability measure*, or the *historical probability measure*. (Some other probability measures will also be used.)

Let $\mathcal{F}_t$ be the filtration generated by (S_t, ρ_t).

Definition 3.19 *Let $\mathbf{P}_* : \mathcal{F} \to [0, 1]$ be a probability measure such that the process $\tilde{S}_t$ is a martingale under $\mathbf{P}_*$ with respect to the filtration $\mathcal{F}_t$. Then $\mathbf{P}_*$ is said to be a risk-neutral probability measure for the bond–stock market (3.4). $\mathbf{E}_*$ denotes the corresponding expectation.*

In particular, $\mathbf{E}_*\{\tilde{S}_\tau | \tilde{S}_1, \ldots, \tilde{S}_t\} = \tilde{S}_t$ for all $\tau > t$.

In literature, a risk-neutral measure is also called a martingale measure.

Proposition 3.20

$$\mathbf{E}_*\{\xi_t|\mathcal{F}_k\} = 0 \quad \forall k, t\colon 0 \le k < t \le T. \tag{3.9}$$

Proof. Let $k = t - 1$. We have that

$$\mathbf{E}_*\left\{\tilde{S}_t|\mathcal{F}_{t-1}\right\} = \tilde{S}_{t-1}\left(1 + \mathbf{E}_*\left\{\xi_t|\mathcal{F}_{t-1}\right\}\right) = \tilde{S}_{t-1}.$$

Then the proof follows for $k = t - 1$, and, therefore, for all $k < t$. □

Proposition 3.21 *The following statements are equivalent.*

(i) *A measure $\mathbf{P}_*$ is risk neutral;*
(ii) $\mathbf{E}_*\left\{\tilde{S}_t|\tilde{S}_{t-1}, \tilde{S}_{t-2}, \dots, \tilde{S}_0\right\} = \tilde{S}_{t-1} \quad \forall t;$
(iii) $\mathbf{E}_*\left\{\tilde{S}_t|\xi_{t-1}, \xi_{t-2}, \dots, \xi_1\right\} = \tilde{S}_{t-1} \quad \forall t;$
(iv) $\mathbf{E}_*\left\{\xi_t|\xi_{t-1}, \dots, \xi_1\right\} = 0 \quad \forall t.$

Proof. We have that $\tilde{S}_t = \tilde{S}_0 \prod_{k=1}^{t}(1 + \xi_k)$, then equivalency of (i) and (ii)–(iii) follows. Further, equivalency of (i) and (iv) follows from the equation $\mathbf{E}_*\left\{\tilde{S}_t|\xi_{t-1}, \xi_{t-2}, \dots, \xi_1\right\} = \tilde{S}_t\left(1 + \mathbf{E}_*\left\{\xi_t|\xi_{t-1}, \dots, \xi_1\right\}\right)$. □

In addition, it follows that if ξ_t does not depend on $\xi_{t-1}, \dots, \xi_1$ under $\mathbf{P}_*$, then the measure $\mathbf{P}_*$ is risk-neutral iff $\mathbf{E}_*\xi_t = 0$ ($\forall t = 1, \dots, n$).

Theorem 3.22 *For any admissible self-financing strategy, the corresponding discounted wealth $\tilde{X}_t$ is a martingale with respect to $\mathcal{F}_t$ under a risk-neutral measure $\mathbf{P}_*$.*

Proof. Let γ_t be the quantity of stock portfolio. Clearly,

$$\tilde{X}_T = X_0 + \mathbf{E}_* \sum_{t=0}^{T-1} \gamma_t\left(\tilde{S}_{t+1} - \tilde{S}_t\right) = \tilde{X}_s + \sum_{t=s}^{T-1} \mathbf{E}_*\gamma_t\tilde{S}_t\xi_{t+1}.$$

Hence

$$\begin{aligned}
\tilde{\mathbf{E}}_*\{X_T \mid \mathcal{F}_s\} &= \tilde{X}_s + \sum_{t=s}^{T-1} \mathbf{E}_*\{\gamma_t\tilde{S}_t\xi_{t+1} \mid \mathcal{F}_s\} \\
&= \tilde{X}_s + \sum_{t=s}^{T-1} \mathbf{E}_*\{\mathbf{E}_*\{\gamma_t\tilde{S}_t\xi_{t+1} \mid \mathcal{F}_t\} \mid \mathcal{F}_s\} \\
&= \tilde{X}_s + \sum_{t=s}^{T-1} \mathbf{E}_*\{\gamma_t\tilde{S}_t\mathbf{E}_*\{\xi_{t+1} \mid \mathcal{F}_t\} \mid \mathcal{F}_s\} = \tilde{X}_s.
\end{aligned}$$

We have used here the fact that $(\gamma_t, \tilde{S}_t)$ is a function of $\{(\rho_k, S_k)\}_{k=1}^t$ (i.e., the vector $(\gamma_t, \tilde{S}_t)$ is $\mathcal{F}_t$-measurable), and the fact that $\mathbf{E}_*\{\xi_{t+1}|\mathcal{F}_t\} = 0$, by Proposition 3.20. This completes the proof. □

3.6 Replicating strategies

Let an integer $T > 0$ be given. Let ψ be an $\mathcal{F}_T$-measurable random variable. (As we know, there exists a deterministic function $F : \mathbf{R}^{2T+2} \to \mathbf{R}$ such that $\psi = F(S_\cdot, \rho_\cdot)$, i.e., $\psi = F(S_0, S_1, \ldots, S_T, \rho_0, \rho_1, \ldots, \rho_T)$).

Definition 3.23 *Let the initial wealth X_0 be given, and let a self-financing strategy (β_t, γ_t) be such that $X_T = \psi$ a.s. for the corresponding wealth. Then the claim ψ is called replicable (attainable, redundant), and the strategy is said to be a replicating strategy (with respect to this claim).*

Definition 3.24 *Let the initial wealth X_0 be given, and let a self-financing strategy (β_t, γ_t) be such that $X_T \geq \psi$ a.s. for the corresponding wealth. Then the strategy is said to be a super-replicating strategy.*

Theorem 3.25 *Let the initial wealth X_0 and a self-financing strategy $\{(\beta_t, \gamma_t)\}$ be such that $X_T = \psi$ a.s. for the corresponding wealth. Let $\mathbf{P}_*$ be a risk-neutral measure, and let $\mathbf{E}_*$ be the corresponding expectation. Let $\mathbf{E}_*\psi^2 < +\infty$. Then*

$$X_0 = \mathbf{E}_* \left(\prod_{t=1}^{T} \rho_t \right)^{-1} \psi.$$

Proof. Clearly, $X_T = \psi$ iff $\tilde{X}_T = \left(\prod_{t=1}^{T} \rho_t \right)^{-1} \psi$ a.s. By Theorem 3.22, it follows that the process $\tilde{X}_t$ is a martingale under $\mathbf{P}_*$ with respect to the filtration $\mathcal{F}_t$, i.e., $\tilde{X}_t = \mathbf{E}_*\{\tilde{X}_T \mid \mathcal{F}_t\}$ for all t. In particular, $X_0 = \tilde{X}_0 = \mathbf{E}_*\tilde{X}_T$. This completes the proof. □

We have not yet referred to the original probability distribution of the process $\{(S_t, \rho_t)\}$ (i.e., of the process $\{(\xi_t, \rho_t)\}$). All previous speculations did not use the original probability measure $\mathbf{P}$; we used only the risk-neutral measure $\mathbf{P}_*$ which is an artificial object; it was not related to the real market. Any particular market model (3.4) is defined by the distribution (or evolution law) for (ξ_t, ρ_t). Clearly, we cannot study a particular market model without taking into account the distribution of $\{(S_t, \rho_t)\}$, i.e., the original probability measure $\mathbf{P}$. The following definition addresses the measure $\mathbf{P}$ for the first time.

Definition 3.26 *If a risk-neutral probability measure $\mathbf{P}_*$ is equivalent to the original measure $\mathbf{P}$, then we call it an equivalent risk-neutral measure.*

First application: the uniqueness of the replicating strategy

Theorem 3.27 *Let the market model be such that there exists an equivalent risk-neutral probability measure $\mathbf{P}_*$. Let a claim ψ be replicable for some initial wealth X_0 and some self-financing strategy $\{(\beta_t, \gamma_t)\}$, i.e., $X_T = \psi$ a.s. for the corresponding wealth. Let $\mathbf{E}_*\psi < +\infty$. Then the initial wealth X_0 is uniquely defined. Moreover, the processes X_t and $\xi_{t+1}\gamma_t$ are uniquely defined up to equivalency. If $\xi_t \neq 0$ a.s. for all t, and the replicating strategy and the corresponding wealth process X_t are uniquely defined up to equivalency.*

Proof. Let the initial wealth $X_0^{(i)}$ and the strategy $\{(\beta_t^{(i)}, \gamma_t^{(i)})\}$ be such that $X_T^{(i)} = \psi$ a.s. for the corresponding wealth $X_t^{(i)}$, $i = 1, 2$.

Let $\tilde{X}^{(i)}$ be the corresponding discounted wealth. By Theorem 3.22, it follows that the processes $\tilde{X}_t^{(i)}$ are martingales under $\mathbf{P}_*$ with respect to the filtration $\mathcal{F}_t$, i.e., $\tilde{X}_t^{(i)} = \mathbf{E}_*\{\tilde{X}_T^{(i)} \mid \mathcal{F}_t\}$. Since $\tilde{X}_T^{(1)} = \tilde{X}_T^{(2)}$, it follows that

$$\tilde{X}_t^{(1)} = \tilde{X}_t^{(2)} \quad \text{a.s.} \quad t = 0, 1, \dots, T.$$

Further, $\tilde{X}_{t+1}^{(i)} - \tilde{X}_t^{(i)} = \gamma_t^{(i)}\tilde{S}_t\xi_{t+1}$, and $\tilde{X}_{t+1}^{(1)} - \tilde{X}_t^{(1)} = \tilde{X}_{t+1}^{(2)} - \tilde{X}_t^{(2)}$ a.s. Hence

$$\gamma_t^{(1)}\tilde{S}_t\xi_{t+1} = \gamma_t^{(2)}\tilde{S}_t\xi_{t+1} \quad \text{a.s.} \quad t = 0, 1, 2, \dots, T-1.$$

Finally, it follows that if $\xi_t \neq 0$ a.s. for all t, then $\gamma_t^{(1)} = \gamma_t^{(2)}$ a.s. for all t. Hence $(\beta_t^{(1)}, \gamma_t^{(1)}) = (\beta_t^{(2)}, \gamma_t^{(2)})$ a.s. for all t. □

3.7 Arbitrage possibilities and arbitrage-free market

Arbitrage is a possibility of a risk-free positive gain. Let us define it formally.

Definition 3.28 *Let $T > 0$ be given. Let $\{(\beta_t, \gamma_t)\}$ be an admissible self-financing strategy, let $\tilde{X}_T$ be the corresponding discounted wealth. If*

$$\mathbf{P}(\tilde{X}_T \geq X_0) = 1, \quad \mathbf{P}(\tilde{X}_T > X_0) > 0, \tag{3.10}$$

then this strategy is said to be an arbitrage strategy. If there exists an arbitrage strategy, then we say that the market model allows an arbitrage.

In fact, some arbitrage possibilities may occasionally exist in real markets, but they cannot last for long. (There is even a special term, 'arbitrageurs', for traders who look for arbitrage.) However, we are interested in models without arbitrage possibilities, since the presence of arbitrage is a sign of some temporary abnormality, and it is used to be corrected by the market forces. Typically, models that allow arbitrage are not useful (unless there is a special interest in arbitrage phenomena). We shall concentrate on arbitrage-free models only.

Problem 3.29 *Let there exist* $t \in \{1, \dots, T\}$ *such that* $\xi_t \geq 0$ *a.s.,* $\mathbf{P}(\xi_t > 0) > 0$. *Prove that this market model allows arbitrage. Hint: take* $\gamma_i = 0$, $i \neq t-1$, *and take* $\gamma_{t-1} > 0$.

Theorem 3.30 *Let a market model be such that there exists a risk-neutral probability measure* $\mathbf{P}_*$ *being equivalent to the original measure* $\mathbf{P}$. *Then the market model does not allow arbitrage.*

Proof. Let $\{(\beta_t, \gamma_t)\}$ be a self-financing admissible strategy that allows arbitrage, i.e., it is such that (3.10) holds for the corresponding discounted wealth. Let there exist a risk-neutral $\mathbf{P}_*$ that is equivalent to the original measure $\mathbf{P}$. In that case,

$$\mathbf{P}_*(\tilde{X}_T \geq X_0) = \mathbf{1}, \quad \mathbf{P}_*(\tilde{X}_T > X_0) > 0.$$

Hence

$$\mathbf{E}_*\tilde{X}_T > X_0. \tag{3.11}$$

But

$$\mathbf{E}_*\tilde{X}_T = X_0 + \mathbf{E}_* \sum_{t=0}^{T-1} \gamma_t \left(\tilde{S}_{t+1} - \tilde{S}_t\right) = X_0 + \sum_{t=0}^{T-1} \mathbf{E}_*\gamma_t\tilde{S}_t\xi_{t+1} = X_0.$$

This contradicts (3.11). □

Problem 3.31 Prove that a risk-neutral equivalent probability measure does not exist for Problem 3.29.

Remark 3.32 The opposite statement to the above theorem, '*absence of arbitrage implies the existence of an equivalent risk-neutral measure*', is also valid under some additional requirements on the strategies. The proof of this assertion is beyond the scope of this book. The equivalence relation between the existence of equivalent risk-neutral measure and the absence of (certain types of) arbitrage is called the *fundamental theorem of asset pricing.*

3.8 A case of complete market

Definition 3.33 *Let a market model be such that there exists an equivalent risk-neutral probability measure* $\mathbf{P}_*$ *(i.e., equivalent to the original measure* $\mathbf{P}$*). A market model is said to be complete if any claim* ψ, *such that* ψ *is* $\mathcal{F}_T$*-measurable and* $\mathbf{E}_*\psi^2 < +\infty$, *is replicable with some initial wealth.*

Note that the initial wealth is in fact uniquely defined by ψ and $\mathbf{P}_*$ (see Theorem 3.25).

Theorem 3.34 *If a market model is complete and there exists an equivalent risk-neutral measure, then this measure is unique (as a measure on $\mathcal{F}_T$).*

Proof. Let $A \in \mathcal{F}_T$, and let $\psi \triangleq \left(\prod_{t=1}^T \rho_t\right)^{-1} \mathbb{I}_A$ ($\mathbb{I}_A$ is the indicator function of A). By the assumption, the claim ψ is replicable with some initial wealth X_0^A. By Theorem 3.27, this X_0^A is uniquely defined. By Theorem 3.25, $\mathbf{E}_*\mathbb{I}_A = X_0^A$ for any risk-neutral measure $\mathbf{P}_*$. Therefore, $\mathbf{P}_*$ is uniquely defined on $\mathcal{F}_T$. □

Corollary 3.35 *Let a market model be such that there exists an equivalent risk-neutral probability measure $\mathbf{P}_*$. In addition, let $\{\xi_t\}$ be independent, and let there exist t and $(a, b, c) \in \mathbf{R}^3$ such that $a \neq b$, $b \neq c$, $c \neq a$,*

$$\mathbf{P}(\xi_t \in \{a, b, c\}) = 1, \quad \mathbf{P}(\xi_t = a) > 0, \quad \mathbf{P}(\xi_t = b) > 0, \quad \mathbf{P}(\xi_t = c) > 0.$$

Then the market model is incomplete.

Proof follows from the existence of more than one risk-neutral probability measure $\mathbf{P}_*$ being equivalent to the original measure $\mathbf{P}$ and such that ξ_t are independent under $\mathbf{P}_*$. (See Problems 2.31 and 2.32.) □

Remark 3.36 By Theorem 3.25, if the initial wealth X_0 and a self-financing strategy $\{(\beta_t, \gamma_t)\}$ are such that $X_T = \psi$ a.s. for the corresponding wealth, then

$$X_0 = \mathbf{E}_* \left(\prod_{t=1}^{T} \rho_t\right)^{-1} \psi.$$

Here $\mathbf{P}_*$ is any risk-neutral measure, and $\mathbf{E}_*$ is the corresponding expectation. By the uniqueness of the initial wealth X_0, this expectation does not depend on the choice of the risk-neutral measure, even if there is more than one risk-neutral measure. It is not a contradiction: all expectations $\mathbf{E}_*$ of all replicable claims are indeed uniquely defined and do not depend on the choice of the risk-neutral measure $\mathbf{P}_*$. However, it can happen that claims $\mathbb{I}_A$ for some $A \in \mathcal{F}_T$ are not replicable.

3.9 Cox–Ross–Rubinstein model

The celebrated Cox–Ross–Rubinstein model (Cox *et al.* 1979) is such that $\rho_t \equiv \rho$ are non-random and constant, and ξ_t are mutually independent random variables that have equal distribution and can have only two values, d_1 and d_2, where d_1, d_2 are given real numbers, $-1 < d_1 < 0 < d_2 < 1$.

Risk-neutral measure for the Cox–Ross–Rubinstein model

Proposition 3.37 *A measure* $\mathbf{P}_*$ *is an equivalent risk-neutral measure if and only if* $\{\xi_t\}$ *are independent under* $\mathbf{P}_*$ *and* $\mathbf{E}_*\xi_t = 0$ $(\forall t)$, *i.e.,*

$$\mathbf{P}_*(\xi_t = d_1) = p, \quad \mathbf{P}_p(\xi_t = d_2) = 1 - p \quad \forall t = 1, 2, \dots, T,$$

where p is such that

$$d_1 p + d_2(1 - p) = 0. \tag{3.12}$$

Proof. By Proposition 3.20, $\mathbf{P}_*$ must be such that $\mathbf{E}_*\{\xi_t|\mathcal{F}_{t-1}\} = \mathbf{E}_*\{\xi_t|\xi_{t-1}, \dots, \xi_1\} = 0$ for the corresponding expectation $\mathbf{E}_*$. Hence

$$\begin{aligned}\mathbf{E}_*\{\xi_t|\mathcal{F}_{t-1}\} &= d_1\mathbf{P}_*(\xi_t = d_1|\mathcal{F}_{t-1}) + d_2\mathbf{P}_*(\xi_t = d_2|\mathcal{F}_{t-1}) \\ &= d_1\mathbf{P}_*(\xi_t = d_1|\mathcal{F}_{t-1}) + d_2(1 - \mathbf{P}_*(\xi_t = d_1|\mathcal{F}_{t-1})) \\ &= 0.\end{aligned}$$

Hence $p \triangleq \mathbf{P}_*(\xi_t = d_1|\mathcal{F}_{t-1})$ is uniquely defined from the equation $d_1 p + d_2(1 - p) = 0$. It follows that p does not depend on t and it is non-random together with the value $\mathbf{P}_*(\xi_t = d_2|\mathcal{F}_{t-1}) = 1 - p$. Therefore,

$$\mathbf{P}_*(\xi_t = d_i) = \mathbf{P}_*(\xi_t = d_i|\mathcal{F}_{t-1}), \quad i = 1, 2.$$

It follows that ξ_t does not depend on $\{\xi_1, \dots, \xi_{t-1}\}$ under $\mathbf{P}_*$. □

Completeness of the Cox–Ross–Rubinstein model

Theorem 3.38 *The Cox–Ross–Rubinstein market model is complete.*

Proof. Let $\tilde{F}(\cdot) : \mathbf{R}^{T+1} \to \mathbf{R}$ be an arbitrary function such that $\mathbf{E}_*\tilde{F}(\tilde{S}_\cdot)^2 < +\infty$. To prove the completeness of the market, it suffices to find an admissible strategy such that $X_T = \rho^T\tilde{F}(\tilde{S}_\cdot)$, i.e., $\tilde{X}_T = \tilde{F}(\tilde{S}_\cdot)$ a.s., and $X_0 = \mathbf{E}_*\tilde{F}(\tilde{S}_\cdot)$.

Let

$$Y_t \triangleq \mathbf{E}_*\{\tilde{F}(\tilde{S}_\cdot) \,|\, \mathcal{F}_t\}.$$

Clearly, $\mathcal{F}_t = \sigma(\xi_1, \dots, \xi_t)$, and

$$Y_t = \mathbf{E}_*\{\tilde{F}(\tilde{S}_\cdot) \,|\, \xi_1, \dots, \xi_t\}.$$

Let $V_0 \triangleq \mathbf{E}_*\tilde{F}(\tilde{S}_\cdot)$. Define the functions $V_t(x) \triangleq \mathbf{E}_*\{\tilde{F}(\tilde{S}_\cdot) \,|\, (\xi_1, \dots, \xi_t) = x\}$, $x \in \mathbf{R}^t$, $t = 1, \dots, T$. Clearly,

$$Y_t = V_t(\xi_1, \dots, \xi_t).$$

By Bayes formula,

$$\mathbf{E}\eta = \mathbf{E}\{\eta \,|\, A\}\mathbf{P}(A) + \mathbf{E}\{\eta \,|\, \bar{A}\}\mathbf{P}(\bar{A})$$

for any integrable random variable η and for any event A and $\bar{A} \triangleq \Omega \backslash A$. Here $\mathbf{E}(\cdot \,|\, A)$ is the expectation for the conditional probability measure $\mathbf{P}(\cdot \,|\, A)$.

We can apply this Bayes formula for the events

$$A = \{\xi_{t+1} = d_1\}, \quad \bar{A} = \{\xi_{t+1} = d_2\},$$

and for the probability space $\Big(\Omega, \mathcal{F}, \mathbf{P}_*(\cdot \,|\, (\xi_1, \ldots, \xi_t) = x)\Big)$. We have that

$$\begin{aligned} V_t(x) &= p\mathbf{E}_*\{\tilde{F}(\tilde{S}_\cdot) \,|\, (\xi_1, \ldots, \xi_t) = x,\ \xi_{t+1} = d_1\} \\ &\quad + (1-p)\mathbf{E}_*\{\tilde{F}(\tilde{S}_\cdot) \,|\, (\xi_1, \ldots, \xi_t) = x,\ \xi_{t+1} = d_2\} \\ &= pV_{t+1}\big((x, d_1)\big) + (1-p)V_{t+1}\big((x, d_2)\big). \end{aligned}$$

Hence

$$\begin{aligned} Y_{t+1} - Y_t &= V_{t+1}(\xi_1, \ldots, \xi_{t+1}) - V_t(\xi_1, \ldots, \xi_t) \\ &= V_{t+1}(\xi_1, \ldots, \xi_{t+1}) - pV_{t+1}(\xi_1, \ldots, \xi_t, d_1) - (1-p)V_{t+1}(\xi_1, \ldots, \xi_t, d_2). \end{aligned}$$

Let $\xi_{t+1} = d_1$, then

$$\begin{aligned} Y_{t+1} - Y_t &= V_{t+1}(\xi_1, \ldots, \xi_t, d_1) - pV_{t+1}(\xi_1, \ldots, \xi_t, d_1) - (1-p)V_{t+1}(\xi_1, \ldots, \xi_t, d_2) \\ &= (1-p)V_{t+1}(\xi_1, \ldots, \xi_t, d_1) - (1-p)V_{t+1}(\xi_1, \ldots, \xi_t, d_2). \end{aligned}$$

Let $\xi_{t+1} = d_2$, then

$$\begin{aligned} Y_{t+1} - Y_t &= V_{t+1}(\xi_1, \ldots, \xi_t, d_2) - pV_{t+1}(\xi_1, \ldots, \xi_t, d_1) - (1-p)V_{t+1}(\xi_1, \ldots, \xi_t, d_2) \\ &= pV_{t+1}(\xi_1, \ldots, \xi_t, d_2) - pV_{t+1}(\xi_1, \ldots, \xi_t, d_1). \end{aligned}$$

In both cases, we have that

$$Y_{t+1} - Y_t = \frac{1}{d_2 - d_1}[V_{t+1}(\xi_1, \ldots, \xi_t, d_2) - \tfrac{1}{2}V_{t+1}(\xi_1, \ldots, \xi_t, d_1)]\xi_{t+1},$$

since (3.12) implies that

$$-\frac{d_1}{d_2 - d_1} = (1-p), \qquad \frac{d_2}{d_2 - d_1} = p.$$

Remember that $\xi_{t+1}\tilde{S}_t = \tilde{S}_{t+1} - \tilde{S}_t$. Hence

$$Y_{t+1} - Y_t = \gamma_t(\tilde{S}_{t+1} - \tilde{S}_t),$$

where

$$\gamma_t = \frac{1}{2\varepsilon\tilde{S}_t}[V_{t+1}(\xi_1,\dots,\xi_t,d_2) - V_{t+1}(\xi_1,\dots,\xi_t,d_1)].$$

Hence Y_t is the discounted wealth which corresponds to the stock portfolio $\{\gamma_t\}$. We have that $Y_T = \tilde{F}(\tilde{S}_\cdot)$. Hence $X_t = \rho^t Y_t$ is the corresponding wealth that replicates the claim $\rho^T F(\tilde{S}_\cdot)$. This completes the proof. □

Remark 3.39 It can be seen that this proof can be extended for the case when ρ is non-random, and, for any t, there exists $\mathcal{F}_{t-1}$-measurable random variables $d_1(t)$ and $d_2(t)$ such that $-1 < d_1(t) < 0 < d_2(t) < 1$ a.s. and $\mathbf{P}(\xi_t \in \{d_1(t), d_2(t)\} \mid \mathcal{F}_{t-1}) = 1$ a.s. It follows that the discrete time market model is also complete for this case of conditionally two-point distribution of ξ_t. Technically, this model is more general than the Cox–Ross–Rubinstein model. It appears that it is the most general assumption that still allows a discrete time market to be complete. (For instance, Corollary 3.35 states that three-point distribution for ξ_t leads to incompleteness.)

Remark 3.40 In fact, the Cox–Ross–Rubinstein model is the most common model in numerical option pricing, since it leads to approximate numerical calculations of option prices via binomial trees, including American options (see Chapter 6). To reduce the number of nodes for binomial trees, it is more convenient to use a model such that

$$\frac{S_t}{S_{t-1}} = \rho(1+\xi_t) \in \{d,u\}, \quad \text{where} \quad du = 1, \quad 0 < d < u,$$

$$\mathbf{P}_*(\rho(1+\xi_t) = u) = p, \quad \mathbf{P}_*(\rho(1+\xi_t) = d) = 1 - p,$$

where

$$p \in (0,1), \quad up + d(1-p) = \rho.$$

This choice of parameters helps to decrease the number of different possible stock prices.

Problem 3.41 (i) Prove that $p = (\rho - d)/(u - d)$ in Remark 3.40.
(ii) Find the (risk-neutral) probability that $S_T = S_0u^2d$ for $t = 3$.

3.10 Option pricing

Options and their types

Let us describe first the most generic options: the European call option and the European put option.

A European call option contract traded (contracted and paid) in $t = 0$ is such that the buyer of the contract has the right (not the obligation) to buy one unit of the underlying asset (from the issuer of the option) in $T > 0$ at the strike price K. The option payoff at time T is $(S_T - K)^+ \triangleq \max(0, S_T - K)$, where S_t is the asset price, and K is the strike price, $t = 0, 1, \dots, T$.

A European put option contract traded in $t = 0$ gives to the buyer of the contract the right to sell one unit of the underlying asset in $T > 0$ at the strike price K. The option payoff (at time T) is $(K - S_T)^+ \triangleq \max(0, K - S_T)$, where S_T is the asset price, and K is the strike price.

We consider more general options. Let an integer $T > 0$ be given.

Definition 3.42 *Let $\psi = F(S_\cdot, \rho_\cdot) = F(S_0, \dots, S_T, \rho_0, \dots, \rho_T)$, where a function $F : \mathbf{R}^{2T+2} \to \mathbf{R}$ is given. The European option with payoff ψ can be defined as a contract traded in $t = 0$ such that the buyer of the contract receives an amount of money equal to ψ at terminal time T.*

The following special cases are covered by this setting:

- (vanilla) European call option: $\psi = (S_T - K)^+$, where $K > 0$ is the strike price;
- (vanilla) European put option: $\psi = (K - S_T)^+$;
- share-or-nothing European call option: $\psi = S_T \mathbb{I}_{\{S_T > K\}}$;
- an Asian option: $\psi = f_1\left(\sum_0^T f_2(S_t)\right)$, where f_i are given functions.

All options in this list are such that payoff time T is given a priori; they are all European options.

An option is called a derivative of the underlying assets. For instance, let the payoff ψ be such that there exists a deterministic function $F : \mathbf{R}^{T+1} \to \mathbf{R}$ such that $\psi = F(S_\cdot)$, i.e., $\psi = F(S_0, S_1, \dots, S_T)$, where S_t are prices of an asset. Then the option with this payoff is a derivative of this asset. European put and call options are derivatives of the underlying stocks (since the payoff does not include ρ_t).

Another important class of options is the class of so-called American options.

Definition 3.43 *Let $F_k(\cdot) : \mathbf{R}^{2k+2} \to \mathbf{R}$ be a given set of functions, $k = 0, 1, \dots, T$. An American option is a contract when the option holder can exercise the option at any time $\tau \in \{0, 1, \dots, T\}$ by his/her choice. In that case, he/she obtains the amount of money equal to $F_\tau(S_0, \dots, S_\tau, \rho_0, \dots, \rho_\tau)$ (or obtains some benefits with this market price).*

For instance, an American put option gives the right (not the obligation) to sell one unit of the underlying asset for a fixed price K (to the issuer of the option); the market value of the payoff is $(K - S_\tau)^+$. Similarly, an American call option gives the right (not the obligation) to sell a unit of the underlying asset (see the detailed discussion in Chapter 6).

Similarly to the case of European options, an American option is said to be a derivative of the underlying assets, if the payoff depends on these assets only. For instance, American put and call options are derivatives of the underlying stock (since the payoff does not include ρ_t).

In addition to the classical American and European put and call options, there are many different types of options that cover different demands of the financial market (we can mention barrier options, lookback options, Bermudian, Israeli, Russian, Parisian, etc.). Most of them can still be classified as either European or American options. However, there are some exceptions: for instance, the Israeli option allows early exercise for the issuer as well as for the holder. Technically, it is not an American option.[1]

Problem 3.44 Let $F_t(\cdot) : \mathbf{R}^{T+1} \to \mathbf{R}$ be a given set of functions, $t = 0, 1, \dots, T$. Consider the following option. The option holder can choose to 'exercise' it at any time $\tau \in \{0, 1, \dots, T\}$. This exercise time is recorded, but the actual payoff is delayed up to time T. At this time T, the option holder obtains the amount of money equal to $F_\tau(S_0, S_1, \dots, S_T)$ (or obtains some benefits with this market price). Does this option belong to any of the classes described here (i.e., European, American, Israeli, Asian, or Irish options)?

(See also Problem 3.70.)

Problem 3.45 Let $F_{s,t}(\cdot) : \mathbf{R}^{T+1} \to \mathbf{R}$ be a given set of functions, where $s, t = 0, 1, \dots, T$, $s \leq t$. Consider the following option. The option holder can choose to 'exercise' it at any time $\tau \in \{0, 1, \dots, T\}$. Moreover, he/she can choose to 'exercise' it for a second time at any time $\theta \in \{\tau, \tau + 1, \dots, T\}$. Any exercise times are recorded, but the actual payoff is delayed up to time T. At this time T, the option holder obtains the amount of money equal to $F_{\tau,\theta}(S_0, S_1, \dots, S_T)$ (or obtains some benefits with this market price). Does this option belong to any of the classes described above?

Remark 3.46 Note that the options described in Problems 3.44 and 3.45 may have economical sense. For instance, take $F_\tau(S_0, S_1, \dots, S_T) = \sum_{t=\tau}^{T}(S_t - K)$, and consider an electricity market, where S_t is the price of an electricity unit at time t. The corresponding option from Problem 3.44 gives the right to enter at any time τ a contract for buying an electricity unit every time $t = \tau, \tau + 1, \dots, T$ for a fixed price K; once started, the contract will continue up to time T without a possibility of early exit. A modification of this option with a possibility of early exit can be represented as a special case of the option from Problem 3.45.

1 The author of this book suggested recently one more modification of the American option: the holder of this option can exercise it at any time by his/her choice; in addition, the holder can retract later the decision to exercise. (The author called it the Irish option, because this research was conducted at the University of Limerick, Ireland.)

The following problem requires some creativity.

Problem 3.47 Invent a new option that does not belong to any of the classes mentioned here. Preferably, suggest an option that has some economical sense. If possible, suggest a pricing method using the approach described below for European and American options.

Fair price of an option

The key role in mathematical finance belongs to a concept of the 'fair price' of options.

The following definition is a discrete time analogue of the definition introduced by Black and Scholes (1973) for a continuous time market.

Definition 3.48 *The fair price of an option of any type (i.e., European, American, etc.) is the minimal initial wealth such that this wealth can be raised to a wealth such that allows to fulfil the option obligation for any market situation with some admissible strategies.*

Let us assume that a probabilistic concept is accepted. This means that the stock price evolves as a random discrete time process, and a probabilistic measure is fixed.

We now rewrite Definition 3.48 more formally for European options.

Definition 3.49 *The fair price of the European option with payoff ψ is the minimal initial wealth X_0 such that there exists an admissible self-financing strategy $\{(\beta_t, \gamma_t)\}$ such that*

$$X_T \geq \psi \quad a.s.$$

for the corresponding wealth.

For a complete market, Definition 3.49 leads to replication.

Theorem 3.50 *Let a market be complete. Then the fair price c_F of the option from Definition 3.49 is*

$$c_F = \mathbf{E}_* \left(\prod_{t=1}^{T} \rho_t \right)^{-1} \psi,$$

and it is the initial wealth X_0 such that there exists an admissible self-financing strategy such that

$$X_T = \psi \quad a.s.$$

for the corresponding wealth.

Proof. From the completeness of the market, it follows that the replicating strategy exists and the corresponding initial wealth is equal to $\rho^{-T}\mathbf{E}_*\psi$. Let us show that it is the fair price. Let $X_0' < c_F$ be another initial wealth, then $\mathbf{E}_*\tilde{X}_T' = X_0' < c_F = \mathbf{E}_* \left(\prod_{t=1}^T \rho_t\right)^{-1} \psi$ for the corresponding discounted wealth $\tilde{X}_t'$. Hence it cannot be true that $\prod_{t=1}^T \rho_t \tilde{X}_T' \geq \psi$ a.s. □

Remark 3.51 Similarly to Propositions 5.44 and 5.45 below, it can be shown that the fair price introduced above is the only price that does not allow arbitrage opportunities either for the buyer or for the seller of an option.

Corollary 3.52 *Consider the Cox–Ross–Rubinstein model such that* $\xi_t \in \{d_1, d_2\}$, $\rho_t \equiv \rho$. *Let* $\psi = F(S_T)$ *be the payoff, where* $F(\cdot)$ *is a given function. Then the fair price of the option is*

$$c_F = \rho^{-T}\mathbf{E}_* F(S_T) = \rho^{-T} \sum_{k=0}^{T} p_k F(\rho^T S_0 (1+d_1)^k (1+d_2)^{T-k}),$$

where

$$p_k = C_T^k p^k (1-p)^{T-k},$$

and where $p \triangleq P_*(\xi_t = d_1)$ *is such that* $\mathbf{E}_*\xi = 0$, *i.e.,* $pd_1 + (1-p)d_2 = 0$.

Proof. We have that $S_T = \rho^T S_0 (1+d_1)^\nu (1+d_2)^{T-\nu}$, where $\nu = \nu(\omega)$ is the (random) number of the values d_1 in the set of the values of ξ_t, $t = 1, \ldots, T$. Clearly, ν has binomial law under $\mathbf{P}_*$. □

Problem 3.53 *Consider the Cox–Ross–Rubinstein model such that* $\xi_t = \pm\varepsilon$, $B_t \equiv B_{t-1}$, $S_0 = 1$, $\varepsilon = 1/4$, $T = 1$. *Find the fair price of the option with payoff* $\psi = \max(S_T - 1, 0)$.

Solution. We have

$$c_F = \tfrac{1}{2} F\left(S_0\left(1+\tfrac{1}{4}\right)\right) + \tfrac{1}{2} F\left(S_0\left(1-\tfrac{1}{4}\right)\right) = \tfrac{1}{2}\cdot\tfrac{1}{8} + \tfrac{1}{2}\cdot 0 = \tfrac{1}{8}. \quad \square$$

Problem 3.54 Consider the Cox–Ross–Rubinstein model such that $\xi_t = \pm\varepsilon$, $B_t = 1.1 B_{t-1}$, $S_0 = 1$, $\varepsilon = 1/4$, $T = 2$. Find the fair price of the option with payoff $F(S_1, \ldots, S_T) = \max(S_T - 1, 0)$.

Solution. We have

$$c_F = 1.1^{-2}\Big[\tfrac{1}{4} F\left(1.1^2 \cdot S_0 \left(1+\tfrac{1}{4}\right)\left(1+\tfrac{1}{4}\right)\right)$$

$$+\tfrac{1}{2} F\left(1.1^2 \cdot S_0 \left(1-\tfrac{1}{4}\right)\left(1+\tfrac{1}{4}\right)\right) + \tfrac{1}{4} F\left(1.1^2 \cdot S_0 \left(1-\tfrac{1}{4}\right)\left(1-\tfrac{1}{4}\right)\right)\Big]$$

$$= 1.1^{-2}\left[\tfrac{1}{4}F\left(1.1^2\cdot\tfrac{5}{4}\cdot\tfrac{5}{4}\right)+\tfrac{1}{2}F\left(1.1^2\cdot\tfrac{5}{4}\cdot\tfrac{3}{4}\right)+\tfrac{1}{4}F\left(1.1^2\cdot\tfrac{3}{4}\cdot\tfrac{3}{4}\right)\right]$$

$$= 1.1^{-2}\left[\tfrac{1}{4}\max\left(1.21\cdot\tfrac{25}{16}-1,0\right)+\tfrac{1}{2}\max\left(1.21\cdot\tfrac{15}{16}-1,0\right)\right.$$

$$\left.+\tfrac{1}{4}\max\left(1.21\cdot\tfrac{9}{16}-1,0\right)\right]$$

$$= 1.1^{-2}[0.223+0.067+0] = 0.24. \quad \square$$

For incomplete markets, Definition 3.49 leads to super-replication. That is not always meaningful. Therefore, there is another popular approach for incomplete markets.

Definition 3.55 *(mean-variance hedging). The fair price of the option is the initial wealth* X_0 *such that* $\mathbf{E}|X_T - \psi|^2$ *is minimal over all admissible self-financing strategies.*

In many cases, this definition leads to the option price calculated as the expectation under a risk-neutral equivalent measure which needs to be chosen by some optimal way, since a risk-neutral equivalent measure is not unique for an incomplete market.

3.11 Increasing frequency and continuous time limit

In reality, prices may change and be measured very frequently. For instance, prices can be given for every five minutes. Therefore, it is reasonable to consider the case when $T \to +\infty$ and $\xi_t \to 0$ (in certain senses). A large number of trading operations per day and per hour leads to a limit model where prices and portfolio are continuous time processes. (In fact, it is a model where a trader can adjust the portfolio with increasingly high frequency.) Therefore, the corresponding continuous time market model for this limit can be useful.

Let $P(t)$ be a continuous time process that describes a stock price, $t \in [0, \tau]$, and let $\tilde{P}(t) = e^{-rt}P(t)$ be the corresponding discounted price. Here $r > 0$ is the bank interest rate, $\tau > 0$ is given terminal time.

Let us assume first that $\tilde{P}(t)$ is a continuous non-constant function, such that the derivative $d\tilde{P}(t)/dt$ is bounded.

Let $t_0 = 0$, $t_{k+1} = k\Delta$, $k = 0, \ldots, T$, where $\Delta \triangleq \tau/T$, $T = 1, 2, \ldots$. Note that $t_n = \tau$.

Consider discrete time discounted prices $\tilde{S}_k = \tilde{P}(t_k)$, $\rho = \rho_k \equiv e^{r\Delta}$.

Consider the discrete time market model with discounted stock prices $\{\tilde{S}_k\}$ and with the self-financing strategy defined by the stock portfolio $\{\gamma_k\}$, where

$\gamma_k = g(t_k)$, and where $g(t) = M(d\tilde{P}/dt)(t)$, $M > 0$. Let $\tilde{X}_T$ be the corresponding discounted wealth. We have that

$$\tilde{X}_T = X(0) + \sum_{k=0}^{T} \gamma_k(\tilde{S}_{k+1} - \tilde{S}_k) \;\rightarrow\; X_0 + \int_0^{\tau} g(t)\frac{d\tilde{P}}{dt}(t)dt$$

as $\Delta = \max_k |t_{k+1} - t_k| \rightarrow 0$, i.e.,

$$\tilde{X}_T \rightarrow X_0 + M \int_0^{\tau} \left|\frac{d\tilde{P}}{dt}(t)\right|^2 dt.$$

Hence

$$\tilde{X}_T \rightarrow +\infty \quad \text{as} \quad M \rightarrow \infty, \quad \Delta = \max_k |t_{k+1} - t_k| \rightarrow 0.$$

It follows that a market model with differentiable $\tilde{P}(t)$ is non-realistic.

Let us consider a different model such that $\tilde{S}_{k+1} = \tilde{S}_k(1 + \xi_{k+1})$, where

$$\xi_{k+1} \stackrel{\Delta}{=} \frac{\tilde{S}_{k+1}}{\tilde{S}_k} - 1$$

obeys the so-called square root law:[2]

$$\operatorname{Var} \xi_k \sim T^{-1},$$

i.e.,

$$\operatorname{Var} \xi_k = \operatorname{Var}\left(\frac{\tilde{P}(t_{k+1})}{\tilde{P}(t_k)} - 1\right) \sim \text{const.}\, |t_{k+1} - t_k|.$$

Therefore,

$$\xi_k \sim \text{const.} T^{-1/2} \sim \text{const.} \sqrt{t_k - t_{k-1}},$$

i.e.,

$$\tilde{P}(t_{k+1}) - \tilde{P}(t_k) \sim \text{const.}\, \tilde{P}(t_k)\sqrt{t_{k+1} - t_k}.$$

This property of $\tilde{P}(t)$ matches the one for the so-called diffusion (Ito) processes that are non-differentiable (they are studied below). To describe these processes, we need Ito calculus (or stochastic calculus). In fact, the diffusion market model (based on Ito calculus) is the ultimate continuous time model.

2 It was Bachelier (1900) who first discovered that the square root law is a law for evolution of stock prices. In fact, Bachelier's model can be approximated to the discrete time market such that $S_{t+1} = S_t + \xi_{t+1}$, $t = 1, 2, \ldots$, i.e., when $S_t = \xi_0 + \xi_1 + \cdots + \xi_t$. (See also comment in Section 5.9.)

Theorem 3.56 *Consider the Cox–Ross–Rubinstein market model with $T \to +\infty$, and*

$$d_1 = -\varepsilon, \quad d_2 = \varepsilon, \quad \varepsilon = \delta T^{-1/2}.$$

Then the sequence $\{\tilde{S}_t\} = \{\tilde{S}_{\varepsilon,t}\}_{t=1}^T$ is such that $\tilde{S}_T$ converges under $\mathbf{P}_*$ in distribution to the log-normal random variable $S_0 e^{\eta}$, where $\eta \sim N(-\delta^2/2, \delta^2)$. (More precisely, $\mathbf{P}_*(\tilde{S}_T \in D) \to \mathbf{P}(S_0 e^{\eta} \in D)$ for any interval $D \subset \mathbf{R}$.)

Proof. We have that

$$\tilde{S}_T = S_0 \prod_{t=1}^{T} (1 + \xi_t) = S_0 (1+\varepsilon)^{\nu} (1-\varepsilon)^{\nu_-},$$

where ν is the (random) number of positive values $+\varepsilon$ in the set of all values of ξ_t, $t = 1, \dots, T$, and where $\nu_- = T - \nu$. Hence

$$\ln \tilde{S}_T = \ln S_0 + \nu \ln(1+\varepsilon) + \nu_- \ln(1-\varepsilon).$$

We have that

$$\ln(1+\varepsilon) = \varepsilon - \varepsilon^2/2 + O(\varepsilon^3), \quad \ln(1-\varepsilon) = -\varepsilon - \varepsilon^2/2 + O(\varepsilon^3).$$

Here $O(\varepsilon^3)$ is a function such that $O(\varepsilon^3)/\varepsilon^3$ is bounded as $\varepsilon \to 0$.

Remember that $\nu + \nu_- = T$. Hence

$$\nu \ln(1+\varepsilon) + \nu_- \ln(1-\varepsilon) = \nu\varepsilon - \nu_-\varepsilon - T\varepsilon^2/2 + T \cdot O(\varepsilon^3).$$

Here $O(\varepsilon^3)$ is a random variable such that $O(\varepsilon^3)/\varepsilon^3$ is bounded uniformly in $\omega \in \Omega$. Let

$$\alpha_T \stackrel{\Delta}{=} \frac{\nu - \nu_-}{\sqrt{T}}.$$

We have that

$$\ln \tilde{S}_T - \ln S_0 = \nu\varepsilon - \nu_-\varepsilon - T\varepsilon^2/2 + TO(\varepsilon^3) = \alpha_T \sqrt{T}\varepsilon - T\varepsilon^2/2 + T \cdot O(\varepsilon^3)$$

$$= \alpha_T \sqrt{T} \frac{\delta}{\sqrt{T}} - T\frac{\delta^2}{2T} + T \cdot O(T^{-3/2}) = \alpha_T \delta - \tfrac{1}{2}\delta^2 + O(T^{-1/2}).$$

Clearly, $\mathbf{P}_*(\xi_t = -\varepsilon) = \mathbf{P}_*(\xi_t = +\varepsilon) = \frac{1}{2}$, and ν has binomial law under $\mathbf{P}_*$. We have that $\nu - \nu_- = 2\nu - T$, and

$$\alpha_T = \frac{\nu - Tp}{\sqrt{Tp(1-p)}}, \quad \text{where} \quad p = 1/2.$$

By the de Moivre–Laplace theorem, we have that α_T converges under $\mathbf{P}_*$ in distribution to a Gaussian random variable $N(0,1)$, i.e., $\mathbf{P}_*(\alpha_T \in D) \to \mathbf{P}(\zeta \in D)$ for any interval $D \subset \mathbf{R}$, where $\zeta \sim N(0,1)$. Hence $\alpha_T\delta - \frac{1}{2}\delta^2$ converges in distribution under $\mathbf{P}_*$ to $\eta \triangleq \delta\zeta - \delta^2/2 \sim N(-\delta^2/2, \delta^2)$, where $\eta \sim N(-\delta^2/2, \delta^2)$. Then the proof follows. □

Corollary 3.57 *Let $r > 0$ and $\rho = \rho(T)$ be such that $\rho^T \to e^{r\tau}$ as $T \to +\infty$. Under the assumptions of Theorem 3.56 and Corollary 3.52,*

$$\rho^{-T}\mathbf{E}_*F(S_T) \to e^{-r\tau}\mathbf{E}F(e^{r\tau}S_0e^{\eta}) \quad \text{as } T \to +\infty,$$

where

$$e^{-r\tau}\mathbf{E}F(e^{r\tau}S_0e^{\eta}) = e^{-r\tau}\frac{1}{\sqrt{2\pi}}\int_{-\infty}^{+\infty} e^{-\frac{x^2}{2}}F\left(S_0\exp\left[r\tau - \frac{\sigma^2}{2} + \sigma x\right]\right)dx.$$

For the case of call and put options, the limit in the last corollary gives the so-called Black–Scholes price that will be discussed below.

3.12 Optimal portfolio selection

In addition to the pricing problem, let us discuss briefly the problem of optimal portfolio selection. Consider the following portfolio selection problem:

$$\text{Maximize} \quad \mathbf{E}U(\tilde{X}_T) \text{ over self-financing admissible strategies.}$$

Here T is the terminal time, $\tilde{X}_t$ is the discounted wealth at time $t = 0, 1, \dots, T$, and $U(\cdot)$ is a given *utility function* that describes risk preferences.

Let $T = 1$, then the problem can be rewritten as

$$\text{Maximize} \quad \mathbf{E}U(X_0 + \gamma_0 S_0 \xi_1) \quad \text{over} \quad \gamma_0 \in \mathbf{R}. \tag{3.13}$$

A solution γ_0 of problem (3.13) can be found given the probability distribution of ξ_1 and given U.

Let U be a strictly convex function, then it can be shown that if $\operatorname{Var}\xi_1 > 0$ then $\mathbf{E}U(X_0 + \gamma_0 S_0\xi_1) \to +\infty$ as $|\gamma_0| \to +\infty$. Clearly, it is meaningless to estimate the performance of a strategy using this U, since this performance criterion leads to the strategies with infinitely large values of $|\gamma_0|$ which make no sense from a practical point of view. Therefore, the optimality criteria with strictly convex U are not practical. The most popular utility functions U are concave, for instance $U(x) = \ln x$ and $U(x) = \delta^{-1}x^{\delta}$, where $\delta < 1$ and $\delta \neq 0$. Another important example of a concave utility function is $U(x) = kx - \mu x^2$, where $k > 0$ and $\mu > 0$ are some constants. Note that non-concave functions U are also used: for instance, if $U(x) = \mathbb{I}_{\{x \geq K\}}$, then $\mathbf{E}U(\tilde{X}(T)) = \mathbf{P}(\tilde{X}(T) \geq K)$. The optimal strategy for this utility function maximizes the probability that the goal value K is achieved for the discounted wealth (i.e., it solves a *goal-achieving* problem).

Problem 3.58 (mean-variance optimization). Assume that $U(x) = kx - \mu x^2$, where $k > 0$ and $\mu > 0$ are some constants. Find optimal γ_0 explicitly given $\mathbf{E}\xi_1$ and $\operatorname{Var}\xi_1$.

To solve this problem, it suffices to represent the expected utility as $\mathbf{E}U(\tilde{X}_1) = -c_0\gamma_0^2 + c_1\gamma_0 + c_2$, where $c_i \in \mathbf{R}$ are constants, and $c_0 > 0$.

The solution to Problem 3.58 represents the special single-stock case of the celebrated Markowitz optimal portfolio in mean-variance setting (Markowitz 1959) which is widely used in practice for multi-stock markets. With some standard techniques from quadratic optimization, its solution can be used for practically interesting problems with constraints such as $\mathbf{E}X_1 \to \max$, $\operatorname{Var}X_1 \leq$ const., or $\operatorname{Var}X_1 \to \min$, $\mathbf{E}X_1 \geq$ const.

Remark 3.59 The solution of the optimal investment problem for a discrete time market with $T > 1$ is much more difficult. For instance, Markovitz's results for quadratic U were extended for the case of $T > 1$ only recently (Li and Ng, 2000).

3.13 Possible generalizations

The discrete time market model allows some other variants, some of which are described below.

- One can consider an additive model for the stock price, when $S_t = S_0 + \xi_1 + \cdots + \xi_t$. This approach leads to the very similar theory. Increasing frequency leads to a normal distribution of prices and allows $S_t < 0$.
- One can consider a multi-stock market model with N stocks S_{it}, $i = 1, \dots, N$, $N \geq 1$, when $\gamma_t = \{\gamma_{it}\}$ are vectors of dimension N, and when the wealth is $X_t = \beta_t B_t + \sum_i \gamma_{it} S_{it}$. The model with $N > 1$ has different properties compared with the case of $N = 1$. For instance, as far as we know, there are no examples of complete discrete time markets with $N > 1$. Some special effects can be found for $N \to +\infty$ (such as strategies that converge to arbitrage). Note also that the most widely used results in practice for optimal portfolio selection are obtained for the case of single-period multi-stock markets, i.e., with $T = 1$ and $N > 1$ (Markowitz mean-variance setting).
- Transaction costs (brokerage fees), bid–ask gap, gap between lending and borrowing rate, taxes, and dividends, can be included in the condition of self-investment.
- Additional constraint can be imposed on the admissible strategies (for instance, we can consider only strategies without short positions, i.e., with $\gamma_t \geq 0$).
- In fact, we addressed only the so-called 'small investor' setting, when the stock prices are not affected by any strategy. For a model that takes into account the impact of a large investor's behaviour, (ρ_m, S_m) is affected by $\{\gamma_k\}_{k<m}$.

3.14 Conclusions

- A discrete time market model is the most generic one, and it covers any market with time series of prices. Strategies developed for this model can be implemented directly. The discrete time model does not require the theory of stochastic integrals.
- Unfortunately, discrete time models are difficult for theoretical investigations, and their role in mathematical finance is limited. A discrete time market model is complete only for the very special case of a two-point distribution (for the Cox–Ross–Rubinstein model and for a model from Remark 3.39). Therefore, pricing is difficult for the general case. Some useful theorems from continuous time setting are not valid for the general discrete time model. Many problems are still unsolved for discrete time market models (including pricing problems and optimal portfolio selection problems).
- The complete Cox–Ross–Rubinstein model of a discrete time market is the main tool in computational finance, since it leads to the so-called method of binomial trees for calculation of option prices. However, this model is restrictive because of a fixed norm of change of price for every step. Formally, the negative impact of this can be reduced by increasing the frequency, i.e., increasing the number of periods and decreasing the size of $|\xi_k|$. Obviously, this leads to numerical difficulties for the large number of periods.
- Continuous time limit models allow a bigger choice of complete markets and provide more possibilities for theoretical investigations.

3.15 Problems

Discrete time market: self-financing strategies

Problem 3.60 Consider a discrete time market model with free borrowing. Let the stock prices be $S_0 = 1$, $S_1 = 1.3$, $S_2 = 1.1$. Let a self-financing strategy be such that the number of stock shares is $\gamma_0 = 1$, $\gamma_1 = 1.2$, $\gamma_2 = 1000$. Let the initial wealth be $X_0 = 1$. Find wealth X_1, X_2 and the quantity of cash in a bank account β_t, $t = 0, 1, 2, 3$.

Problem 3.61 Consider a discrete time bond–stock market model. Let the stock prices be $S_0 = 1$, $S_1 = 1.3$, $S_2 = 1.1$. Let the bond prices be $B_0 = 1$, $B_1 = 1.1B_0$, $B_2 = 1.05B_1$. Let a self-financing strategy be such that the number of stock shares is $\gamma_0 = 1$, $\gamma_1 = 1.2$. Let the initial wealth be $X_0 = 1$. Find the wealth X_1, X_2 and the quantity of bonds β_t, $t = 0, 1, 2$.

Problem 3.62 Consider a discrete time bond–stock market with prices from Problem 3.61. Let the initial wealth be $X_0 = 1$. Let a self-financing strategy be such that the number of stock shares at the initial time be $\gamma_0 = 1/2$. Find $\gamma_1, X_1, X_2, \beta_i$, $i = 0, 1, 2$ for the constantly rebalanced portfolio.

Solve Problems 3.11 and 3.12.

Problem 3.63 (Make your own model). Introduce a reasonable version of the discrete time market model that takes into account transaction costs (a brokerage fee), and derive the equation for the wealth evolution for self-financing strategy here. (*Hint:* transaction costs may be per transaction, or may be proportional to the size of transaction or may be of a mixed type.)

Discrete time market: arbitrage and completeness

Solve Problem 3.29.

Problem 3.64 Prove that an equivalent risk-neutral probability measure does not exist for Problem 3.29.

Problem 3.65 Let a market model be such that $\rho_t \equiv \rho$, where ρ is non-random and given, $\{\xi_t\}$ are independent, and let there exist $(a, b, c) \in \mathbf{R}^3$ such that $\mathbf{P}(\xi_t \in \{a, b, c\}) = 1$ for all t. Explain in which cases the market is arbitrage-free, allows arbitrage, complete or incomplete:

(i) $\rho = 1, a = b = 0.1, c = -0.05,$
$\mathbf{P}(\xi_t \in \{a, c\}) = 1, \mathbf{P}(\xi_t = a) \in (0, 1);$

(ii) $\rho = 1.1, a = b = 0.15, c = -0.05,$
$\mathbf{P}(\xi_t \in \{a, c\}) = 1, \mathbf{P}(\xi_t = a) \in (0, 1);$

(iii) $\rho = 1.1, a = b = 0.15, c = 1.1,$
$\mathbf{P}(\xi_t \in \{a, c\}) = 1, \mathbf{P}(\xi_t = a) \in (0, 1);$

(iv) $\rho = 1.1, a = b = 0, c = -0.05,$
$\mathbf{P}(\xi_t \in \{a, c\}) = 1, \mathbf{P}(\xi_t = a) \in (0, 1);$

(v) $\rho = 1.1, a = 0.05, b = 0.1, c = -0.05,$
$\mathbf{P}(\xi_t = a) > 0, \mathbf{P}(\xi_t = b) > 0, \mathbf{P}(\xi_t = c) > 0;$

(vi) $\rho = 1.1, a = -0.05, b = 0.15, c = -0.05,$
$\mathbf{P}(\xi_t = a) > 0, \mathbf{P}(\xi_t = b) > 0, \mathbf{P}(\xi_t = c) > 0.$

Problem 3.66 Let a market model be such that $\rho_t \equiv \rho$, where ρ is non-random and given, $\{\xi_t\}$ are independent, and let there exist $(a, b, c) \in \mathbf{R}^3$ such that $\mathbf{P}(\eta_t \in \{a, b, c\}) = 1$ for all t, where $\eta_t \triangleq S_{t+1}/S_t$. Explain in which cases the market is arbitrage-free, allows arbitrage, complete or incomplete:

(i) $\rho = 1, a = b = 1.1, c = 0.95,$
$\mathbf{P}(\eta_t \in \{a, c\}) = 1, \mathbf{P}(\eta_t = a) \in (0, 1);$

(ii) $\rho = 1.1, a = b = 1.15, c = 0.95,$
$\mathbf{P}(\eta_t \in \{a, c\}) = 1, \mathbf{P}(\eta_t = a) \in (0, 1);$

(iii) $\rho = 1.1, a = b = 1.15, c = 1.1,$

$\mathbf{P}(\eta_t \in \{a, c\}) = 1, \mathbf{P}(\eta_t = a) \in (0, 1);$

(iv) $\rho = 1.1, a = b = 1, c = 0.95,$

$\mathbf{P}(\eta_t \in \{a, c\}) = 1, \mathbf{P}(\eta_t = a) \in (0, 1);$

(v) $\rho = 1.1, a = 1.05, b = 1.1, c = 0.95,$

$\mathbf{P}(\eta_t = a) > 0, \mathbf{P}(\eta_t = b) > 0, \mathbf{P}(\eta_t = c) > 0;$

(vi) $\rho = 1.1, a = 0.95, b = 1.15, c = 0.95,$

$\mathbf{P}(\eta_t = a) > 0, \mathbf{P}(\eta_t = b) > 0, \mathbf{P}(\eta_t = c) > 0.$

Problem 3.67 Let $a, b \in \mathbf{R}$ be given such that $a \leq b$, $\mathbf{P}(S_t/S_{t-1} \in \{a, b\}) = 1$, $\mathbf{P}(S_t/S_{t-1} = a) > 0$, and $\mathbf{P}(S_t/S_{t-1} = b) > 0$ for all t. In addition, let $B_t = 1.07 \cdot B_{t-1}$ for all t. Find conditions on a and b such that the market is arbitrage-free.

Option price for the Cox–Ross–Rubinstein model

Problem 3.68 Consider the Cox–Ross–Rubinstein model such that $\tilde{S}_t = \tilde{S}_{t-1}(1+\xi_t)$, $\xi_t = \pm\varepsilon$. Let $B_t \equiv B_{t-1}$, $S_0 = 1$, $F(S_\cdot) = \max(S_T - 1.1, 0)$, $\varepsilon = 1/5$, $T = 1$. Find the fair price of the option.

Problem 3.69 Consider the Cox–Ross–Rubinstein model such that $\tilde{S}_t = \tilde{S}_{t-1}(1+\xi_t)$, $\xi_t = \pm 1/4$. Let $B_t = 1.1 B_{t-1}$, $S_0 = 1$, $F(S_\cdot) = \max(S_T - 1.2, 0)$, $T = 2$. Find the fair price of the option.

Challenging problem

Problem 3.70 Consider the option described in Problem 3.44. Prove that there exists an American option (Definition 3.43) such that its fair price is equal to the fair price of the option from Problem 3.44.

4 Basics of Ito calculus and stochastic analysis

This chapter introduces the stochastic integral, stochastic differential equations, and core results of Ito calculus.

4.1 Wiener process (Brownian motion)

Let $T > 0$ be given, $t \in [0, T]$.

Definition 4.1 *We say that a continuous time random process $w(t)$ is a (one-dimensional) Wiener process (or Brownian motion) if*

(i) $w(0) = 0$;
(ii) $w(t)$ *is Gaussian with* $\mathbf{E}w(t) = 0$, $\mathbf{E}w(t)^2 = t$, *i.e.,* $w(t)$ *is distributed as* $N(0, t)$;
(iii) $w(t + \tau) - w(t)$ *does not depend on* $\{w(s),\ s \le t\}$ *for all* $t \ge 0$, $\tau > 0$.

Theorem 4.2 *(N. Wiener). There exists a probability space* $(\Omega, \mathcal{F}, \mathbf{P})$ *such that there exists a pathwise continuous process with these properties.*

This is why we call it the Wiener process. The corresponding set Ω in Wiener's proof of this theorem is the set $C(0, T)$. Remember that $C(0, T)$ denotes the set of all continuous functions $f : [0, T] \to \mathbf{R}$.

Corollary 4.3 *Let* $\Delta t > 0$, $\Delta w(t) \triangleq w(t + \Delta t) - w(t)$, *then* $\operatorname{Var} \Delta w = \Delta t$.

Corollary 4.4

$$\mathbf{E}\left(\frac{\Delta w(t)}{\Delta t}\right)^2 = \frac{1}{\Delta t}.$$

This can be interpreted as

$$\frac{\Delta w(t)}{\Delta t} \sim \frac{1}{\sqrt{\Delta t}} \quad \text{as} \quad \Delta t \to 0.$$

This means that a Wiener process cannot have pathwise differentiable trajectories. Its trajectories are very irregular (but they are still continuous a.s.).

Let us list some basic properties of $w(t)$:

- sample paths maintain continuity;
- paths are non-differentiable;
- paths are not absolutely continuous, and any path of the process $(t, w(t))$ is a *fractal* line in $\mathbf{R}^2$, or a very irregular set.

Definition 4.5 *We say that a continuous time process* $w(t) = (w_1(t), \dots, w_n(t)) : [0, +\infty) \times \Omega \to \mathbf{R}^n$ *is a (standard) n-dimensional Wiener process if*

(i) $w_i(t)$ *is a (one-dimensional) Wiener process for any* $i = 1, \dots, n$*;*
(ii) the processes $\{w_i(t)\}$ *are mutually independent.*

Remark 4.6 Let $A = \{a_{ij}\} \in \mathbf{R}^{n \times n}$ be a matrix such that $\sum_j a_{ij}^2 = 1$. Then the process $\tilde{w}(t) \stackrel{\Delta}{=} Aw(t)$ is also said to be a Wiener process (but not a standard Wiener process, since it has correlated components).

We shall omit the word 'standard' below; all Wiener processes in this book are assumed to be standard.

For simplicity, one can assume for the first reading that $n = 1$, and all processes used in this chapter are one-dimensional. After that, one can read this chapter again taking into account the general case.

Proposition 4.7 *A Wiener process is a Markov process.*

Proof. We consider an n-dimensional Wiener process $w(t)$. Let $\mathcal{F}_t^w$ be the filtration generated by $w(t)$. We have that $w(t + \tau) = w(t + \tau) - w(t) + w(t)$. Further, let times $\{t_i\}$ and s be such that $t_i > s$, $i = 1, \dots, k$. Clearly, $w(s)$ is $\mathcal{F}_s^w$-measurable and does not depend on $\{w(t_i) - w(s)\}_{i=1}^k$. For any bounded measurable function $F : \mathbf{R}^{nk} \to \mathbf{R}$, we have that

$$
\begin{aligned}
&\mathbf{E}\{F(w(t_1), w(t_2), \dots, w(t_k)) \,|\, \mathcal{F}_s^w\} \\
&\quad = \mathbf{E}\{F_1(w(t_1) - w(s), w(t_2) - w(s), \dots, w(t_k) - w(s), w(s)) \,|\, \mathcal{F}_s^w\} \\
&\quad = F_2(w(s))
\end{aligned}
$$

for some measurable functions $F_1 : \mathbf{R}^{nk} \to \mathbf{R}$ and $F_2 : \mathbf{R}^n \to \mathbf{R}$. It follows that $w(s)$ is a Markov process. □

Proposition 4.8 *Let* $\mathcal{F}_t$ *be a filtration such that an n-dimensional Wiener process* $w(t)$ *is adapted to* $\mathcal{F}_t$*, and* $w(t + \tau) - w(t)$ *does not depend on* $\mathcal{F}_t$*. Then* $w(t)$ *is a martingale with respect to* $\mathcal{F}_t$*.*

Proof. We have that $w(t+\tau) = w(t+\tau) - w(t) + w(t)$. Hence

$$\begin{aligned}\mathbf{E}\{w(t+\tau) \,|\, \mathcal{F}_t\} &= \mathbf{E}\{w(t+\tau) - w(t) \,|\, \mathcal{F}_t\} + w(t) \\ &= \mathbf{E}(w(t+\tau) - w(t)) + w(t) = w(t),\end{aligned}$$

since $w(t+\tau) - w(t)$ does not depend on $\mathcal{F}_t$. Therefore, the martingale property holds. □

Corollary 4.9 *A Wiener process $w(t)$ is a martingale. (In other words, if $\mathcal{F}_t^w$ is the filtration generated by $w(t)$, then $w(t)$ is a martingale with respect to $\mathcal{F}_t^w$.)*

Up to the end of this chapter, we assume that we are given an n-dimensional Wiener process $w(t)$ and the filtration $\mathcal{F}_t$ such as described in Proposition 4.8. One may assume that this filtration is generated by the process $(w(t), \eta(t))$, where $\eta(s)$ is a process independent from $w(\cdot)$. We assume also that $t \in [0, T]$, where $T > 0$ is given deterministic terminal time.

4.2 Stochastic integral (Ito integral)

Stochastic integral for step functions

Let $w(t)$ be a one-dimensional Wiener process. Repeat that $\mathcal{F}_t$ is a filtration such as described in Proposition 4.8.

Notation: Let $\mathcal{L}_{22}^0$ be the set of $\mathcal{F}_t$-adapted functions $f(t,\omega)$ such that there exists an integer $N > 0$, a set of times $0 = t_0 < t_1 < \cdots < t_N = T$, and a sequence $\{\xi_k\}_{k=1}^N \subset \mathcal{L}_2(\Omega, \mathcal{F}, P)$, such that $f(t) = \xi_k$ for $t \in [t_k, t_{k+1})$, $k = 0, \dots, N-1$.

Clearly, all these functions are pathwise step functions.

Problem 4.10 Prove that, in the definition above, ξ_k are $\mathcal{F}_{t_k}$-measurable.

Definition 4.11 *Let $f(\cdot) \in \mathcal{L}_{22}^0$. The value*

$$I(f) \triangleq \sum_{k=0}^{N-1} f(t_k)[w(t_{k+1}) - w(t_k)]$$

is said to be the Ito integral of f, or stochastic integral, and it is denoted as $\int_0^T f(t)dw(t)$, i.e.,

$$\int_0^T f(t)dw(t) = I(f).$$

Theorem 4.12 *Let* $f, g \in \mathcal{L}_{22}^0$. *Then*

(i) $\mathbf{E}\int_0^T f(t)dw(t) = 0$;

(ii) $\mathbf{E}\left(\int_0^T f(t)dw(t)\right)^2 = \mathbf{E}\int_0^T |f(t)|^2 dt$;

(iii) $\mathbf{E}\int_0^T f(t)dw(t)\int_0^T g(t)dw(t) = \mathbf{E}\int_0^T f(t)g(t)dt$.

Proof is straightforward and follows from the definitions given above.

Theorem 4.13 *Let* $f \in \mathcal{L}_{22}^0$. *Then* $\mathbf{E}\{\int_0^T f(t)dw(t) \mid \mathcal{F}_t\} = \int_0^t f(s)dw(s)$.

Cauchy sequences in $\mathcal{L}_2$

First, let us describe some properties of random variables from $\mathcal{L}_2(\Omega, \mathcal{F}, \mathbf{P})$.

Definition 4.14 *Let* $\xi \in \mathcal{L}_2(\Omega, \mathcal{F}, \mathbf{P})$, *and let* $\{\xi_k\} \subset \mathcal{L}_2(\Omega, \mathcal{F}, \mathbf{P})$ *be a sequence such that* $\mathbf{E}|\xi_k - \xi|^2 \to 0$. *Then we say that this sequence converges to* ξ *in* $L_2(\Omega, \mathcal{F}, \mathbf{P})$ *(or* $\xi_k \to \xi$ *as* $k \to +\infty$ *in* $L_2(\Omega, \mathcal{F}, \mathbf{P})$, *or* $\xi = \lim \xi_k$ *in* $L_2(\Omega, \mathcal{F}, \mathbf{P})$*).*

Remember that $L_2(\Omega, \mathcal{F}, \mathbf{P})$ is the set of classes of **P**-equivalent random variables from $\mathcal{L}_2(\Omega, \mathcal{F}, \mathbf{P})$.[1]

Definition 4.15 *Let* $\{\xi_k\} \subset \mathcal{L}_2(\Omega, \mathcal{F}, \mathbf{P})$ *be a sequence such that, for any* $\varepsilon > 0$, *there exists* $N > 0$ *such that* $\mathbf{E}|\xi_k - \xi_m|^2 < \varepsilon$ *for all* k *and* m *such that* $k > N$, $m > N$. *Then we call this sequence a Cauchy sequence in* $L_2(\Omega, \mathcal{F}, \mathbf{P})$.

Theorem 4.16 *(i) Any Cauchy sequence* $\{\xi_k\}$ *in* $L_2(\Omega, \mathcal{F}, \mathbf{P})$ *has a unique limit in* $L_2(\Omega, \mathcal{F}, \mathbf{P})$. *In other words, there exists a unique (up to* **P***-equivalency) element* $\xi \in \mathcal{L}_2(\Omega, \mathcal{F}, \mathbf{P})$ *such that* $\xi_i \to \xi$ *in* $L_2(\Omega, \mathcal{F}, \mathbf{P})$.

(ii) Let $\xi \in \mathcal{L}_2(\Omega, \mathcal{F}, \mathbf{P})$, *and let* $\{\xi_k\} \subset \mathcal{L}_2(\Omega, \mathcal{F}, \mathbf{P})$ *be a sequence. If* $\xi_k \to \xi$ *in* $L_2(\Omega, \mathcal{F}, \mathbf{P})$, *then* $\mathbf{E}\xi_k \to \mathbf{E}\xi$ *and* $\mathbf{E}|\xi_k|^2 \to \mathbf{E}|\xi|^2$.

(iii) Let $\xi, \eta \in \mathcal{L}_2(\Omega, \mathcal{F}, \mathbf{P})$, *and let* $\{\xi_k\} \subset \mathcal{L}_2(\Omega, \mathcal{F}, \mathbf{P})$, $\{\eta_k\} \subset \mathcal{L}_2(\Omega, \mathcal{F}, \mathbf{P})$ *be some sequences. If* $\xi_k \to \xi$ *and* $\eta_k \to \eta$ *in* $L_2(\Omega, \mathcal{F}, \mathbf{P})$, *then* $\mathbf{E}\xi_k\eta_k \to \mathbf{E}\xi\eta$.

Ito integral for general functions

Notation: We denote by $\mathcal{L}_{22}$ the set of all random processes that can be approximated by processes from $\mathcal{L}_{22}^0$ in the following sense: for any $f \in \mathcal{L}_{22}^0$, there exists a sequence $\{f_k(\cdot)\}_{k=1}^{+\infty} \subset \mathcal{L}_{22}^0$ such that $\mathbf{E}\int_0^T |f(t) - f_k(t)|^2 dt \to 0$ as $k \to +\infty$.

1 We do not need to refer to the definition of Banach and Hilbert spaces and their properties that are usually studied in functional analysis or function spaces courses. However, it may be useful to note that $L_2(\Omega, \mathcal{F}, \mathbf{P})$ is a Banach space and a Hilbert space with the norm $\|\xi\| = (\mathbf{E}|\xi|^2)^{1/2}$.

Note that all processes from $\mathcal{L}_{22}$ are adapted to the filtration $\mathcal{F}_t$ (more precisely, if $f \in \mathcal{L}_{22}$, then $f(t)$ is $\mathcal{F}_t$-measurable for a.e. (almost every) t.[2]

Theorem 4.17 *Let $f \in \mathcal{L}_{22}$, and let $\{f_k(\cdot)\}_{k=1}^{+\infty} \subset \mathcal{L}_{22}^0$ be such that $\mathbf{E}\int_0^T |f(t) - f_k(t)|^2 dt \to 0$ as $k \to +\infty$. Then $\{I(f_k)\}_{k=1}^{\infty}$ is a Cauchy sequence in $L_2(\Omega, \mathcal{F}, \mathbf{P})$, where $I(f_k) = \int_0^T f_k(t)dw(t)$. This sequence converges in $L_2(\Omega, \mathcal{F}, \mathbf{P})$, and its limit depends only on f and does not depend on the choice of the approximating sequence (in the sense that all possible modifications of the limit are $\mathbf{P}$-equivalent).*

Definition 4.18 *The limit of $I(f_k)$ in $\mathcal{L}_2(\Omega, \mathcal{F}, \mathbf{P})$ from the theorem above is said to be the Ito integral (stochastic integral)*

$$\int_0^T f(t)dw(t).$$

Theorem 4.19 *Let $f, g \in \mathcal{L}_{22}$. Then*

(i) $\mathbf{E}\int_0^T f(t)dw(t) = 0;$

(ii) $\mathbf{E}\left(\int_0^T f(t)dw(t)\right)^2 = \mathbf{E}\int_0^T |f(t)|^2 dt;$

(iii) $\mathbf{E}\int_0^T f(t)dw(t) \int_0^T g(t)dw(t) = \mathbf{E}\int_0^T f(t)g(t)dt.$

Proof follows from the properties for approximating functions from $\mathcal{L}_{22}^0$. □

Theorem 4.20 *Let $f, g \in \mathcal{L}_{22}$. Then*

(i) $\mathbf{E}\left\{\int_0^T f(t)dw(t) \,\middle|\, \mathcal{F}_s\right\} = \int_0^s f(t)dw(t);$

(ii) $\mathbf{E}\left\{\int_s^T f(t)dw(t) \,\middle|\, \mathcal{F}_s\right\} = 0;$

(iii)
$$\mathbf{E}\left\{\int_0^T f(t)dw(t) \int_0^T g(t)dw(t) \,\middle|\, \mathcal{F}_s\right\} = \int_0^s f(t)dw(t) \int_0^s g(t)dw(t) + \mathbf{E}\left\{\int_s^T f(t)g(t)dt \,\middle|\, \mathcal{F}_s\right\}.$$

Proof follows again from the properties for approximating functions from $\mathcal{L}_{22}^0$. □

Definition 4.21 *A modification of a process $\xi(t, \omega)$ is any process $\xi'(t, \omega)$ such that $\xi(t, \omega) = \xi'(t, \omega)$ for a.e. t, ω.*

2 In fact, processes $\xi \in \mathcal{L}_{22}$ are measurable as mappings $\xi : [0, T] \times \mathbf{P} \to \mathbf{R}$ with respect to the completion of the σ-algebra generated by all mappings $\xi_0 : [0, T] \times \mathbf{P} \to \mathbf{R}$ such that $\xi_0 \in \mathcal{L}_{22}^0$.

Theorem 4.22 *Let $T > 0$ be fixed, and let $f \in \mathcal{L}_{22}$. Then the process $\int_0^t f(s)dw(s)$ is pathwise continuous in $t \in [0, T]$ (more precisely, there exists a modification of the process $\int_0^t f(s)dw(s)$ that is continuous in $t \in [0, T]$ a.s. (i.e., with probability 1, or for a.e. (almost every) ω).*

Note that

(i) a stochastic integral is defined up to **P**-equivalency;
(ii) it is not defined pathwise, i.e., we cannot construct it as a function of T for a fixed ω.

Ito integral for a random time interval

Let $\mathcal{F}_t$ be the filtration generated by the Wiener process $w(t)$, and let $f \in \mathcal{L}_{22}$. Let τ be a Markov time with respect to $\mathcal{F}_t$, $\tau \in [0, T]$. Then $\mathbb{I}_{\{t\leq\tau\}} f(t) \in \mathcal{L}_{22}$. In that case, we can define the Ito integral for a random time interval $[0, \tau]$ as

$$\int_0^\tau f(t)dw(t) \triangleq \int_0^T \mathbb{I}_{\{t\leq\tau\}} f(t)dw(t).$$

It follows that

$$\mathbf{E}\int_0^\tau f(t)dw(t) = 0.$$

In particular, it can be shown that $w(\tau) = w(0) + \int_0^\tau dw(t)$, hence $\mathbf{E}w(\tau) = 0$. Note that it holds for Markov times τ and may not hold for arbitrary random time τ. For instance, if time τ is such that $w(\tau) = \max_{s\in[0,T]} w(s)$, then τ is not a Markov time and $\mathbf{E}w(\tau) > 0$.

In addition,

$$\mathbf{E}\left\{\int_0^T f(t)dw(t)\middle|\mathcal{F}_\tau\right\} = \int_0^\tau f(t)dw(t).$$

Vector case

Let $w(t)$ be an n-dimensional Wiener process, and let $\mathcal{F}_t$ be a filtration such as described in Proposition 4.8. Let $f = (f_1, \ldots, f_n) : [0, T] \times \Omega \to \mathbf{R}^{1\times n}$ be a (vector row) process such that $f_i \in \mathcal{L}_{22}$ for all i. Then we can define the Ito integral

$$\int_0^T f(t,\omega)dw(t) \triangleq \sum_{i=1}^n \int_0^T f_i(t,\omega)dw_i(t).$$

The right-hand part is well defined by the previous definitions.

Ito processes

Definition 4.23 *Let $w(t)$ be an n-dimensional Wiener process, $\alpha \in \mathcal{L}_{22}$, $a \in \mathcal{L}_2(\Omega, \mathcal{F}_0, \mathbf{P})$. Let a random process $\beta = (\beta_1, \dots, \beta_n)$ take values in $\mathbf{R}^{1\times n}$, and let $\beta_i \in \mathcal{L}_{22}$ for all i. Let*

$$y(t) = a + \int_0^t \alpha(r)dr + \int_0^t \beta(r)dw(r).$$

Then the process $y(t)$ is said to be an Ito process. The expression

$$dy(t) = \alpha(t)dt + \beta(t)dw(t)$$

is said to be the stochastic differential (or Ito differential) of $y(t)$. The process $\alpha(t)$ is said to be the drift coefficient, and $\beta(t)$ is said to be the diffusion coefficient.

Theorem 4.24 *An Ito process*

$$y(t) = a + \int_0^t \alpha(r)dr + \int_0^t \beta(r)dw(r)$$

is a martingale with respect to $\mathcal{F}_t$ if and only if $\alpha(t) \equiv 0$ up to equivalency.

Proof. By Theorem 4.20, it follows that if $\alpha \equiv 0$ then y is a martingale. Proof of the opposite statement needs some analysis. □

4.3 Ito formula

One-dimensional case

Let us assume first that $\alpha, \beta \in \mathcal{L}_{22}$ are one-dimensional processes, and $w(t)$ is a one-dimensional process

$$y(t) = y(s) + \int_s^t \alpha(r)dr + \int_s^t \beta(r)dw(r),$$

i.e., $y(t)$ is an Ito process, and

$$dy(t) = \alpha(t)dt + \beta(t)dw(t).$$

Let $V(\cdot, \cdot) : \mathbf{R} \times [0, T] \to \mathbf{R}$ be a continuous function such that its derivatives V'_t, V'_x, V''_{xx} are continuous (and such that some additional conditions on their growth are satisfied).

Theorem 4.25 *(Ito formula, or Ito lemma). The process $V(y(t),t)$ is also an Ito process, and its stochastic differential is*

$$d_t V(y(t),t) = \frac{\partial V}{\partial t}(y(t),t)dt + \frac{\partial V}{\partial y}(y(t),t)dy(t) + \frac{1}{2}\frac{\partial^2 V}{\partial x^2}(y(t),t)\beta(t)^2 dt. \tag{4.1}$$

Note that the last equation can be rewritten as

$$d_t V(y(t),t) = \left[\frac{\partial V}{\partial t}(y(t),t) + \mathcal{A}(t)V(y(t),t)\right]dt + \frac{\partial V}{\partial y}(y(t),t)\beta(t)dw(t),$$

where $\mathcal{A}(t)$ is the differential operator

$$\mathcal{A}(t)v(x) = \frac{dv}{dx}(x)\alpha(t) + \frac{1}{2}\frac{d^2 v}{dx^2}(x)\beta(t)^2.$$

Remark 4.26 In fact, the formula for the drift and diffusion coefficients of the process $V(y(t),t)$ was first obtained by A.N. Kolmogorov as long ago as 1931[3] for the special case when $y(t)$ is a Markov (diffusion) process. It gives (4.1) for this case (see Shiryaev (1999), p. 263, where it was outlined that it would be natural to call it the Kolmogorov–Ito formula).

Proof of Theorem 4.25 is based on the Taylor series and the estimate

$$\Delta y \triangleq y(t+\Delta t) - y(t) \sim a(t)\Delta t + \beta(t)\Delta w,$$

where $(\Delta w)^2 \sim \Delta t$. □

Example 4.27 Let $y(t) = w(t)^2$, then $dy(t) = 2w(t)dw(t) + dt$.

Let $\alpha_i, \beta_i \in \mathcal{L}_{22}$,

$$dy_i(t) = \alpha_i(t)dt + \beta_i(t)dw(t), \quad i = 1,2.$$

Theorem 4.28 *Let $y(t) \triangleq y_1(t)y_2(t)$, then*

$$dy(t) = y_1(t)dy_2(t) + y_2(t)dy_1(t) + \beta_1(t)\beta_2(t)dt.$$

3 *Mathematische Annalen* 104 (1931), 415–458.

The vector case

Let us assume first that $w(t)$ is an n-dimensional Wiener process. Let random processes $a = (a_1, \dots, a_m)$ and $\beta = \{\beta_{ij}\}$ take values in $\mathbf{R}^m$ and $\mathbf{R}^{m\times n}$ respectively, and let $a_i \in \mathcal{L}_{22}$ and $b_{ij} \in \mathcal{L}_{22}$ for all i, j. Let $y(t)$ be an m-dimensional Ito process, and

$$dy(t) = \alpha(t)dt + \beta(t)dw(t),$$

i.e.,

$$dy(t) = \alpha(t)dt + \sum_{i=1}^{n} \beta_i(t)dw_i(t).$$

Here β_i are the columns of the matrix β. (It is an equation for a vector process that has not been formally introduced before; we simply require that the corresponding equation holds for any component). Let $V(\cdot,\cdot) : \mathbf{R}^m \times [0, T] \to \mathbf{R}$ be a continuous function such that the derivatives V_t', V_x', V_{xx}'' are continuous (and such that some additional conditions on their growth are satisfied). Note that V_x' takes values in $\mathbf{R}^{1\times m}$, and V_{xx}'' takes values in $\mathbf{R}^{m\times m}$.

Theorem 4.29 *(Ito formula for the vector case). The process $V(y(t), t)$ is also an Ito process, and its stochastic differential is*

$$d_t V(y(t), t) = \frac{\partial V}{\partial t}(y(t), t)dt + \frac{\partial V}{\partial y}(y(t), t)dy(t)$$
$$+ \frac{1}{2}\sum_{i=1}^{m} \beta_i(t)^{\top} \frac{\partial^2 V}{\partial x^2}(y(t), t)\beta_i(t)dt.$$

Note that the last equation can be rewritten as

$$d_t V(y(t), t) = \left[\frac{\partial V}{\partial t}(y(t), t) + \mathcal{A}(t)V(y(t), t)\right]dt + \frac{\partial V}{\partial y}(y(t), t)\beta(t)dw(t),$$

where $\mathcal{A}(t)$ is the differential operator

$$\mathcal{A}(t)v(x) = \frac{dv}{dx}(x)\alpha(t) + \frac{1}{2}\sum_{i=1}^{m} \beta_i(t)^{\top}\frac{d^2 v}{dx^2}(x)\beta_i(t).$$

In addition, it can be useful to note that

$$\sum_{i=1}^{m} \beta_i^{\top}\frac{\partial^2 V}{\partial x^2}\beta_i \equiv \operatorname{Tr}\left[\beta\beta^{\top}\frac{\partial^2 V}{\partial x^2}\right],$$

where Tr denotes the *trace* of a matrix (i.e., the summa of all eigenvalues).

Problem 4.30 Prove that Theorem 4.28 follows from Theorem 4.29.

4.4 Stochastic differential equations (Ito equations)

4.4.1 Definitions

Let $f(x, t, \omega) : \mathbf{R}^m \times [0, T] \times \Omega \to \mathbf{R}^m$ and $b(x, t, \omega) : \mathbf{R}^m \times [0, T] \times \Omega \to \mathbf{R}^{m \times n}$ be some functions. Let the processes $f(x, t, \omega)$ and $b(x, t, \omega)$ be adapted to the filtration $\mathcal{F}_t$ for all x.

Definition 4.31 *Let $a = (a_1, \dots, a_m)$ be a random vector with values in $\mathbf{R}^m$ such that $a_i \in \mathcal{L}_2(\Omega, \mathcal{F}_0, \mathbf{P})$. Let an m-dimensional process $y(t) = y_1(t), \dots, y_m(t)$ be such that $y_i \in \mathcal{L}_{22}$ and*

$$y(t) = a + \int_0^t f(y(r), r, \omega)dr + \int_0^t b(y(r), r, \omega)dw(r) \quad \text{for all } t \quad \text{a.s.}$$

We say that the process $y(t)$, $t \in [0, T]$, is a solution of the stochastic differential equation (Ito equation)

$$\begin{cases} dy(t) = f(y(t), t, \omega)dt + b(y(t), t, \omega)dw(t), \\ y(0) = a. \end{cases} \tag{4.2}$$

Let $\mathcal{L}_{22}(s, T)$ be the set of functions $f : [s, T] \times \Omega \to \mathbf{R}$ defined similarly to $\mathcal{L}_{22}$.

Definition 4.32 *Let $s \in [0, T]$, and let $a = (a_1, \dots, a_m)$ be a random vector with values in $\mathbf{R}^m$ such that $a_i \in \mathcal{L}_2(\Omega, \mathcal{F}_s, \mathbf{P})$. Let an m-dimensional process $y(t) = y_1(t), \dots, y_m(t)$ be such that $y_i \in \mathcal{L}_{22}(s, T)$ and*

$$y(t) = a + \int_s^t f(y(r), r, \omega)dr + \int_s^t b(y(r), r, \omega)dw(r) \quad \forall t \quad \text{a.s.} \tag{4.3}$$

We say that the process $y(t)$, $t \in [s, T]$, is a solution of the stochastic differential equation (Ito equation)

$$\begin{cases} dy(t) = f(y(t), t, \omega)dt + b(y(t), t, \omega)dw(t), \\ y(s) = a. \end{cases} \tag{4.4}$$

Problem 4.33 Does it make a difference if one requires that (4.3) holds a.s. for all t (instead of 'for all t a.s.')?

Example 4.34 The following result is immediate. Let $\alpha, \beta \in \mathcal{L}_{22}$, $a \in \mathcal{L}_2(\Omega, \mathcal{F}_s, \mathbf{P})$. The equation

$$\begin{cases} dy(t) = \alpha(t)dt + \beta(t)dw(t), \\ y(s) = a \end{cases} \tag{4.5}$$

has a solution

$$y(t) = a + \int_s^t \alpha(r)dr + \int_s^t \beta(r)dw(r), \quad t \in [s, T].$$

Remark 4.35 For the case when $f(x,t) : \mathbf{R}^m \times [0,T] \to \mathbf{R}^m$ and $b(x,t) : \mathbf{R}^m \times [0,T] \to \mathbf{R}^{m\times n}$ are non-random, the solution $y(t)$ of equation (4.2) is a Markov process. In that case, it is called a *diffusion process.*

For the general case of random f or b, the process $y(t)$ is not a Markov process; in that case, it is sometimes called a *diffusion-type process* (but not a diffusion process).

In particular, if $n = m = 1, f(x,t) \equiv ax, b(x,t) \equiv \sigma x$, then the equation for $y(t)$ is the equation for the stock price $dS(t) = S(t)[adt + \sigma dw(t)]$, which we will discuss below.

4.4.2 The existence and uniqueness theorem

Theorem 4.36 *(The existence and uniqueness theorem). Let (random) functions* $f(x,t,\omega) : \mathbf{R}^m \times [0,T] \times \Omega \to \mathbf{R}^m$, $b(x,t,\omega) : \mathbf{R}^m \times [0,T] \times \Omega \to \mathbf{R}^{m\times n}$ *be continuous in* (x,t) *with probability 1. Further, let the processes* $f(x,\cdot)$ *and* $b(x,\cdot)$ *be* $\mathcal{F}_t$*-adapted for all* x*, and let there exist a constant* $C > 0$ *such that*

$$|f(x,t,\omega)| + |b(x,t,\omega)| \le C(|x| + 1),$$

$$|f(x,t,\omega) - f(x_1,t,\omega)| + |b(x,t,\omega) - b(x_1,t,\omega)| \le C|x - x_1|$$

for all $x, x_1 \in \mathbf{R}$, $t \in [0,t]$, *a.s. Let* $s \in [0,T]$, *and let* $a \in \mathcal{L}_2(\Omega, \mathcal{F}_s, \mathbf{P})$. *Then equation (4.4) has a unique solution* $y \in \mathcal{L}_{22}(s,T)$ *(unique up to equivalency).*

Here and below $|x| = \left(\sum_{i=1}^m x_i^2\right)^{1/2}$ denotes the Euclidean norm for $x \in \mathbf{R}^m$, and $|x| = \left(\sum_{i,j=1}^m x_{ij}^2\right)^{1/2}$ denotes the Frobenius matrix norm for $x \in \mathbf{R}^{m\times m}$.

We shall deal mostly with Ito equations with known solutions, when it can be verified that the Ito equation is satisfied. However, a question arises over whether this solution is unique. Therefore, for our purposes, it is more important to prove the uniqueness claimed in the theorem. Let us prove the uniqueness only.

Proof of Theorem 4.36 (uniqueness). Let $y_i(t)$ be two solutions. We have that

$$y_1(t) - y_2(t) = \int_0^t [f(y_1(s),s) - f(y_2(s),s)]ds$$
$$+ \int_0^t [\beta(y_1(s),s) - \beta(y_2(s),s)]dw(s).$$

Hence

$$
\begin{aligned}
&\mathbf{E}|y_1(t)-y_2(t)|^2\\
&\quad=\mathbf{E}\left|\int_0^t [f(y_1(s),s)-f(y_2(s),s)]ds+\int_0^t [\beta(y_1(s),s)-\beta(y_2(s),s)]dw(s)\right|^2\\
&\quad\leq 2\mathbf{E}\left|\int_0^t [f(y_1(s),s)-f(y_2(s),s)]ds\right]^2\\
&\qquad+2\mathbf{E}\left[\int_0^t [\beta(y_1(s),s)-\beta(y_2(s),s)]dw(s)\right|^2\\
&\quad\leq 2\mathbf{E}\left|\int_0^t |f(y_1(s),s)-f(y_2(s),s)|ds\right]^2\\
&\qquad+2\int_0^t \mathbf{E}[\beta(y_1(s),s)-\beta(y_2(s),s)|^2ds\\
&\quad\leq 2C^2\mathbf{E}\left|\int_0^t |y_1(s)-y_2(s)|ds\right|^2+2C^2\int_0^t \mathbf{E}|y_1(s)-y_2(s)|^2ds\\
&\quad\leq 2C^2T\mathbf{E}\int_0^t |y_1(s)-y_2(s)|^2ds+2C^2\int_0^t \mathbf{E}|y_1(s)-y_2(s)|^2ds.
\end{aligned}
$$

We have used here the inequality $(a+b)^2 \leq 2a^2+2b^2$, and the inequality

$$\int_0^t |g(s)|ds \leq \sqrt{t}\left[\int_0^t |g(s)|^2ds\right]^{1/2}$$

that holds for all square integrable functions $g : [0,t] \to \mathbf{R}$.

Proposition 4.37 *(Bellman inequality). Let $T > 0$ and $k_i \geq 0$ be given, $i = 1,2$. Then there exists $C > 0$ such that $\sup_{y\in[0,T]} m(t) \leq Ck_2$ for any function $m(\cdot) : [0,T] \to \mathbf{R}$ such that*

$$0 \leq m(t) \leq k_1 \int_0^t m(s)ds + k_2.$$

Let $z(t) \triangleq y_1(t) - y_2(t)$ and $m(t) \triangleq \mathbf{E}|z(t)|^2$.

We have that

$$0 \leq m(t) \leq 2C^2(T+1)\int_0^t m(s)ds.$$

By Bellman inequality, it follows that $0 \leq m(t) \leq 0$, i.e., $m(t) \equiv 0$, i.e., $y_1(t) = y_2(t)$ a.s. for all t. This completes the proof of the uniqueness in Theorem 4.36. □

Note that the solution of the Ito equation (4.4) is not defined backward (i.e., for $t < s$); in other words, the Cauchy condition $y(s) = a$ cannot be imposed at the *end* of the time interval. This is different from the case of ordinary differential equations, where the simple change of time variable from t to $-t$ makes forward and backward equations mutually interchangeable.

4.4.3 Continuous time white noise

Sometimes, especially in engineering literature, the Ito equation appears in the form

$$\begin{cases} \dfrac{dy}{dt}(t) = f(y(t), t) + b(y(t), t)\dfrac{dw}{dt}(t), \\ y(s) = a. \end{cases} \tag{4.6}$$

This way is legitimate, provided that integral equation (4.3) is assumed. In that case, we do not need to give an interpretation for $[dw/dt](t)$. (Remember that the process $w(t)$ is non-differentiable, and $y(t)$ is also non-differentiable.) Alternatively, the derivative dw/dt can be defined in a class of so-called generalized random processes (constructed similarly to the generalized deterministic functions such as the delta function). This generalized process dw/dt is a continuous time analogue of the discrete time white noise. This approach is used mainly for the case of linear equations with constant b in control system theory.

4.4.4 Examples of explicit solutions for Ito equations

In Problems 4.38, 4.39, and 4.41 below, we assume that $n = m = 1$.

Processes with log-normal distributions

Problem 4.38 Let $a, \sigma \in \mathbf{R}$, $y_0 \in \mathbf{R}$. Show that the equation

$$\begin{cases} dy(t) = ay(t)dt + \sigma y(t)dw(t), \\ y(0) = y_0 \end{cases} \tag{4.7}$$

has the unique solution

$$y(t) = y_0 \exp\left(at - \frac{\sigma^2}{2}t + \sigma w(t)\right), \quad t \geq 0.$$

Hint 1: For the uniqueness, use the existence and uniqueness Theorem 4.36.

Hint 2: Apply the Ito formula for $y(t)$. For instance, set $V(x, t) = e^x$, $\xi(t) = \ln y_0 + at - (\sigma^2/2)/t + \sigma w(t)$. Then the process $y(t) = V(\xi(t), t)$ is such that $y(0) = y_0$. The Ito formula should be used to verify that the stochastic differential equation is satisfied for the process $y(t) = V(\xi(t), t)$.

Problem 4.39 Let $a, \sigma \in \mathbf{R}$, $y_s \in \mathcal{L}_2(\Omega, \mathcal{F}_s, \mathbf{P})$. Show that the equation

$$\begin{cases} dy(t) = ay(t)dt + \sigma y(t)dw(t), \\ y(s) = y_s \end{cases} \tag{4.8}$$

has the unique solution

$$y(t) = y_s \exp\left(\left[a - \frac{\sigma^2}{2}\right](t-s) + \sigma(w(t) - w(s))\right), \quad t \geq s.$$

Note that the solutions of equations (4.7) and (4.8) are distributed log-normally conditionally given y_s.

A generalization

Problem 4.40 Let $w(t)$ be an n-dimensional Wiener process, $\alpha(\cdot) \in \mathcal{L}_{22}$, let $\sigma(t) = (\sigma_1(t), \dots, \sigma_n(t))$ be a process with values in $\mathbf{R}^{1\times n}$ such that $\sigma_i(\cdot) \in \mathcal{L}_{22}$, and let some conditions on the growth for a, σ be satisfied (it suffices to assume that they are bounded). Let $y_s \in \mathcal{L}_2(\Omega, \mathcal{F}_s, \mathbf{P})$. Show that the equation

$$\begin{cases} dy(t) = a(t)y(t)dt + y(t)\sigma(t)dw(t), \\ y(s) = y_s \end{cases} \tag{4.9}$$

has the unique solution

$$y(t) = y_s \exp\left(\int_s^t a(r)dr - \frac{1}{2}\int_s^t |\sigma(r)|^2 dr + \int_s^t \sigma(r)dw(r)\right), \quad t \geq s.$$

Ornstein–Uhlenbek process

Problem 4.41 Let $\alpha, \lambda, \sigma \in \mathbf{R}$, $y_s \in \mathcal{L}_2(\Omega, \mathcal{F}_s, \mathbf{P})$. Show that the equation

$$\begin{cases} dy(t) = (\alpha - \lambda y(t))dt + \sigma dw(t), \\ y(s) = y_s \end{cases} \tag{4.10}$$

has a unique solution

$$y(t) = e^{-\lambda t} y_s + \int_s^t e^{-\lambda(t-r)}\alpha dr + \int_s^t e^{-\lambda(t-r)}\sigma dw(r), \quad t \geq s.$$

Hint: use that

$$\int_s^t e^{-\lambda(t-r)}\sigma dw(r) = e^{-\lambda t}\int_s^t e^{\lambda r}\sigma dw(r).$$

If $\lambda > 0$, then the solution $y(t)$ of (4.10) is said to be an Ornstein–Uhlenbek process. This process converges (in a certain sense) to a stationary Gaussian process as $t \to +\infty$ (continuous time stationary processes; this convergency will be discussed in Chapter 9).

4.5 Diffusion Markov processes and Kolmogorov equations

One-dimensional case

Let (non-random) functions $f : \mathbf{R} \times [0, T] \to \mathbf{R}$ and $b : \mathbf{R} \times [0, T] \to \mathbf{R}$ be given. Let $y(t)$ be a solution of the stochastic differential equation

$$\begin{cases} dy(t) = f(y(t), t)dt + b(y(t), t)dw(t), & t > s, \\ y(s) = a. \end{cases} \tag{4.11}$$

We shall denote this solution as $y^{a,s}(t)$. As was mentioned above, this process is called a diffusion process; it is a Markov process.

Let functions $\Psi : \mathbf{R} \to \mathbf{R}$ and $\varphi : \mathbf{R} \times [0, T] \to \mathbf{R}$ be such that certain conditions on their smoothness and growth are satisfied (for instance, it suffices to assume that they are continuous and bounded).

Let $V(x, s)$ be the solution of a Cauchy problem for the parabolic equation

$$\begin{aligned} &\frac{\partial V}{\partial s}(x, s) + \mathcal{A}(s)V(x, s) = -\varphi(x, s), \\ &V(x, T) = \Psi(x). \end{aligned} \tag{4.12}$$

Here $x \in \mathbf{R}$, $s \in [0, T]$,

$$\mathcal{A}(t)v(x) = \frac{dv}{dx}(x)f(x, t) + \frac{1}{2}\frac{d^2v}{dx^2}(x)b(x, t)^2.$$

Note that (4.12) is a so-called backward parabolic equation, since the Cauchy condition is imposed at the end of the time interval.

We assume that the functions f, b, Ψ, φ are such that this boundary value problem (4.12) has a unique solution V such that it has continuous derivatives V'_t, V'_x, and V''_{xx}.

Theorem 4.42

$$\Psi(y^{x,s}(T)) + \int_s^T \varphi(y^{x,s}(t), t)dt = V(x, s) + \int_s^T \frac{\partial V}{\partial y}(y^{x,s}(t), t)b(y(t), t)dw(t).$$

Proof. By the Ito formula,

$$
\begin{aligned}
\Psi(y^{x,s}(T)) - V(x,s) &= V(y^{x,s}(T),T) - V(y^{x,s}(s),s) \\
&= \int_s^T \left[\frac{\partial V}{\partial t} + \mathcal{A}V\right](y^{x,s}(t),t)dt + \int_s^T \frac{\partial V}{\partial y}(y^{x,s}(t),t)b(y(t),t)dw(t) \\
&= -\int_s^T \varphi(y^{x,s}(t),t)dt + \int_s^T \frac{\partial V}{\partial y}(y^{x,s}(t),t)b(y(t),t)dw(t).
\end{aligned}
$$

Then the proof follows. □

Vector case

Let $w(t)$ be an n-dimensional Wiener process. Let (non-random) functions $f : \mathbf{R}^m \times [0,T] \to \mathbf{R}^m$ and $b : \mathbf{R}^m \times [0,T] \to \mathbf{R}^{m\times n}$ be given, $a \in \mathbf{R}^m$. Let $y(t)$ be a solution of the stochastic differential equation

$$
\begin{cases} dy(t) = f(y(t),t)dt + b(y(t),t)dw(t), & t > s, \\ y(s) = a. \end{cases} \tag{4.13}
$$

We shall denote this solution as $y^{a,s}(t)$.

Let functions $\Psi : \mathbf{R}^m \to \mathbf{R}$ and $\varphi : \mathbf{R}^m \times [0,T] \to \mathbf{R}$ be such that certain conditions on their smoothness and growth are satisfied (it suffices again to assume that they are continuous and bounded).

Let $V(x,s)$ be the solution of the Cauchy problem for the parabolic equation

$$
\begin{aligned}
&\frac{\partial V}{\partial s}(x,s) + \mathcal{A}(s)V(x,s) = -\varphi(x,s), \\
&V(x,T) = \Psi(x).
\end{aligned} \tag{4.14}
$$

Here $x \in \mathbf{R}^m$, $s \in [0,T]$,

$$
\mathcal{A}(t)v(x) = \frac{dv}{dx}(x)f(x,t) + \frac{1}{2}\sum_{i=1}^{n} b_i(x,t)^\top \frac{d^2v}{dx^2}(x)b_i(x,t).
$$

Here b_i are the columns of the matrix b.

Again, (4.12) is a so-called backward parabolic equation, since the Cauchy condition is imposed at the end of the time interval.

Theorem 4.43

$$
\Psi(y^{x,s}(T)) + \int_s^T \varphi(y^{x,s}(t),t)dt = V(x,s) + \int_s^T \frac{\partial V}{\partial y}(y^{x,s}(t),t)b(y(t),t)dw(t).
$$

Proof repeats the proof of Theorem 4.42. □

The following corollary gives the probabilistic representation of the solution V of the Cauchy problem for the parabolic equation.

Corollary 4.44

$$V(x,s) = \mathbf{E}\Psi(y^{x,s}(T)) + \mathbf{E}\int_s^T \varphi(y^{x,s}(t),t)dt.$$

Case of a bounded domain

The same approach is used for boundary value problems for parabolic equations: the solution can be represented via expectation of a function of a random process. If there is a boundary of the domain, then these functions include first exit time from the domain for the random process.

Let $D \subset \mathbf{R}^m \times [0,T]$ be a domain with the boundary ∂D.

Under the assumptions of Theorem 4.43, let $\tau^{x,s} \triangleq \inf_{t>0}\{(y^{x,s}(t),t) \notin D\}$.

Let the domain D have a regular enough boundary ∂D, and let functions $\psi : D \to \mathbf{R}$ and $\varphi : D \to \mathbf{R}$ be such that certain conditions on their smoothness and growth are satisfied (for instance, it suffices to assume that they are continuous and bounded).

Let $V(x,s)$ be the solution of the boundary problem for the parabolic equation

$$\begin{aligned} &\frac{\partial V}{\partial s}(x,s) + \mathcal{A}(s)V(x,s) = -\varphi(x,s), \\ &V(x,s)|_{(x,s)\in\partial D} = \psi(x,s). \end{aligned} \tag{4.15}$$

Theorem 4.45

$$\begin{aligned} &\psi(y^{x,s}(\tau^{x,s}),\tau^{x,s}) + \int_s^{\tau^{x,s}} \varphi(y^{x,s}(t),t)dt \\ &\quad = V(x,s) + \int_s^{\tau^{x,s}} \frac{\partial V}{\partial y}(y^{x,s}(t),t)b(y(t),t)dw(t). \end{aligned}$$

Proof is again similar to the proof of Theorem 4.42. □

The following corollary gives the probabilistic representation of V.

Corollary 4.46

$$V(x,s) = \mathbf{E}\psi(y^{x,s}(\tau^{x,s}),\tau^{x,s}) + \mathbf{E}\int_s^{\tau^{x,s}} \varphi(y^{x,s}(t),t)dt.$$

Remark 4.47 A similar approach can be used for Dirichlet problems for elliptic equations: their solution can be represented via functions of diffusion processes with time-independent coefficients and with infinite time horizon $T = +\infty$.

Some terminology

- The differential operator $\mathcal{A}$ is said to be the differential operator generated by the process $y(t)$.
- Equation (4.12) is said to be the backward Kolmogorov (parabolic) equation for the process $y(t)$ (or Kolmogorov–Fokker–Planck equation). Historically, diffusion Markov processes were studied via these equations before the appearance of the Ito calculus. The novelty of the Ito calculus was that it gave a very powerful method that covers very general settings, in particular non-Markov processes.
- The equation for the probability density function of $y(t)$ is called the forward Kolmogorov equation (forward Kolmogorov–Fokker–Planck equation). It is the so-called *adjoint equation* for equation (4.12) and it can be derived from (4.12).

For the examples below, we assume that $n = m = 1$.

Example 4.48 Let $y(t) = y^{x,s}(t)$ be a solution of the stochastic differential equation

$$\begin{cases} dy(t) = ay(t)dt + \sigma y(t)dw(t), & t > s, \\ y(s) = x. \end{cases}$$

Here $a \in \mathbf{R}$, $\sigma \in \mathbf{R}$ are given. Then the function $u(x,s) = \mathbf{E}\Psi(y^{x,s}(T))$ can be represented as the solution of the Cauchy problem for the parabolic equation

$$\frac{\partial u}{\partial t}(x,t) + ax\frac{\partial u}{\partial x}(x,t) + \frac{1}{2}\sigma^2 x^2 \frac{\partial^2 u}{\partial x^2}(x,t) = 0, \quad x \in \mathbf{R},\ t < T,$$

$$u(x,T) = \Psi(x).$$

It suffices to apply Theorem 4.42 with $f(x,t) \equiv ax$, $b(x,t) \equiv \sigma x$, and the corresponding operator is

$$\mathcal{A}(t)u = ax\frac{du}{dx} + \frac{1}{2}\sigma^2 x^2 \frac{d^2u}{dx^2}. \quad \square$$

Example 4.49 The Cauchy problem for the *heat equation* (heat parabolic equation) is

$$\frac{\partial u}{\partial t}(x,t) + \frac{1}{2}\frac{\partial^2 u}{\partial x^2}(x,t) = 0, \quad x \in \mathbf{R},\ t < T,$$

$$u(x,T) = \Psi(x)$$

allows solution

$$u(x,s) = \mathbf{E}\Psi(y^{x,s}(T)) = \mathbf{E}\Psi(w(T) - w(s) + x) = \mathbf{E}\Psi(\eta), \tag{4.16}$$

where $\eta = w(T) - w(s) + x$ is Gaussian with law $N(x, T - s)$. It suffices to apply Theorem 4.42 with $f \equiv 0$, $b \equiv 1$, then $y^{x,s}(t) = w(t) - w(s) + x$, and the corresponding operator is $\mathcal{A}(t) = (1/2)(d^2/dx^2)$. □

In Example 4.49, representation (4.16) is said to be the probabilistic representation of the solution. In particular, it follows that

$$u(x,s) = \int_{-\infty}^{+\infty} p(x,y,s,T)\Psi(y),$$

where

$$p(x,y,s,T) = \frac{1}{\sqrt{2(T-s)}} e^{-\frac{|x-y|^2}{2(T-s)}}$$

is the probability density function for $N(x, T - s)$. Note that this function is also well known in the theory of parabolic equations: it is the so-called fundamental solution of the heat equation.

The representation of functions of the stochastic processes via solution of parabolic partial differential equations (PDEs) helps to study stochastic processes: one can use numerical methods developed for PDEs (i.e., finite differences, fundamental solutions, etc.).

On the other hand, the probabilistic representation of a solution of parabolic PDEs can also help to study PDEs. For instance, one can use Monte Carlo simulation for numerical solution of PDEs. Some theoretical results can also be proved easier with probabilistic representation (for example, the so-called *maximum principle for parabolic equations* follows from this representation: if $\varphi \geq 0$ and $\Psi \geq 0$ in (4.15), then $V \geq 0$).

Remark 4.50 It follows that the diffusion process $y(t)$ can be considered as the characteristics of the parabolic equation, by an analogy with the first-order hyperbolic equations (the case of $b \equiv 0$). It is known from physical models that the propagation described by the first-order hyperbolic equations has a bounded speed, and that the speed of heat propagation is infinite, i.e., the 'physical' diffusion process has unlimited speed. This fact is linked with non-differentiability of $y(t)$.

4.6 Martingale representation theorem

In this section, we assume that $w(t)$ is an n-dimensional vector process, $\mathcal{F}_t$ is the filtration generated by $w(t)$, *and a wider filtration is not allowed.*

The following result is known as the Clark theorem or Clark–Hausmann–Ocone theorem.

Theorem 4.51 *Let $\xi \in L_2(\Omega, \mathcal{F}_T, \mathbf{P})$. Then there exists an n-dimensional process $f = (f_1, \dots, f_n)$ with values in $\mathbf{R}^{1\times n}$ such that $f_i \in \mathcal{L}_{22}$ for all i and*

$$\xi = \mathbf{E}\xi + \int_0^T f(t)dw(t).$$

Proof. (a) Let us consider first the case when $\xi = \prod_{k=1}^m \eta_k$, where $\eta_k \triangleq g_k(w(t_k))$, $0 \le t_m < \dots < t_1 \le T$, and where $g_k : \mathbf{R}^n \to \mathbf{R}$ are some measurable bounded functions, $k = 1, \dots, n$. Let $m = 1$, then the theorem's statement follows from Theorem 4.43 applied for $\Psi(x) = g_1(x)$, $T = t_1$, and $y^{0,0}(t) = w(t)$.

For $m > 1$, we use the induction by m. Therefore, it suffices to show that if the theorem holds for $m - 1$, then it implies that it holds for m. Let $\hat{\eta}_{m-1} \triangleq \prod_{k=1}^{m-1} \eta_k$. Let us assume that there exists an n-dimensional process $f_{m-1} = (f_{m-1,1}, \dots, f_{m-1,n})$ with values in $\mathbf{R}^{1\times n}$ such that $f_{m-1,i} \in \mathcal{L}_{22}$ for all i and

$$\hat{\eta}_{m-1} = \mathbf{E}\hat{\eta}_{m-1} + \int_0^T f_{m-1}(t)dw(t).$$

(It is the induction assumption.) Clearly, $\xi = \eta_m \hat{\eta}_{m-1}$, and

$$\begin{aligned}\xi &= \mathbf{E}\{\xi \mid \mathcal{F}_{t_m}\} + \xi - \mathbf{E}\{\xi \mid \mathcal{F}_{t_m}\}\\ &= \eta_m \mathbf{E}\{\hat{\eta}_{m-1} \mid \mathcal{F}_{t_m}\} + \eta_m(\hat{\eta}_{m-1} - \mathbf{E}\{\hat{\eta}_{m-1} \mid \mathcal{F}_{t_m}\})\\ &= \eta_m \mathbf{E}\{\hat{\eta}_{m-1} \mid \mathcal{F}_{t_m}\} + \eta_m \int_{t_m}^T f_{m-1}(t)dw(t).\end{aligned}$$

By the Markov property of $w(t)$, it follows that there exists a measurable bounded function $\hat{\phi} : \mathbf{R} \to \mathbf{R}$ such that $\mathbf{E}\{\hat{\eta}_{m-1} \mid \mathcal{F}_{t_m}\} = \hat{\phi}(w(t_m))$. Hence $\eta_m \mathbf{E}\{\hat{\eta}_{m-1} \mid \mathcal{F}_{t_m}\} = \phi(w(t_m))$, where $\phi(x) \triangleq g_m(x)\hat{\phi}(x)$. It follows that

$$\xi = \phi(w(t_m)) + \int_{t_m}^T \eta_m f_{m-1}(t)dw(t).$$

By Theorem 4.43 applied for $\Psi(x) = \phi(x)$ and $T = t_m$, we have that there exists an n-dimensional process $\bar{f}(t)$ with values in $\mathbf{R}^{1\times n}$ and with components from $\mathcal{L}_{22}$ such that

$$\phi(w(t_m)) = \mathbf{E}\phi(w(t_m)) + \int_0^{t_m} \tilde{f}(t)dw(t).$$

Then the proof follows for this special ξ.

(b) By the linearity of the Ito integral, the proof follows for all random variables $\xi = \sum_k c_k \mathbb{I}_{A_k}$, with some constants c_k, and with random events $A_k = \cap_{i=1}^{n_k}\{w(t_{ik}) \in J_{ik}\}$, where $n_k \ge 1$, $0 \le t_{i1} < \dots < t_{in_k} \le T$, and where J_{ik} are measurable subsets of $\mathbf{R}^n$. (Note that $\mathbb{I}_{A_k} = \prod_{i=1}^{n_k} \mathbb{I}_{\{w(t_{ik})\in J_{ik}\}}$, so the theorem statement proved in (a) can be used for $\mathbb{I}_{A_k}$.)

(c) For the case when $\xi \in \mathcal{L}_{22}$ is of the general type, the proof follows from the fact that the set of random variables described in (b) is dense in $L_2(\Omega, \mathcal{F}, \mathbf{P})$. (We omit this part.) □

Note that Theorem 4.51 allows equivalent formulation as the following martingale representation theorem.

Theorem 4.52 *Let $\xi(t)$ be a martingale with respect to the filtration $\mathcal{F}_t$ generated by a Wiener process $w(t)$ such that $\mathbf{E}\xi(T)^2 < +\infty$. Then there exists a process $f(t)$ with values in $\mathbf{R}^{1\times n}$ and with components from $\mathcal{L}_{22}$ such that*

$$\xi(t) = \mathbf{E}\xi(t) + \int_0^t f(s)dw(s), \quad t \in [0, T].$$

Proof. Apply Theorem 4.51 to $\xi(T)$. □

Corollary 4.53 *Any martingale described in Theorem 4.52 is pathwise continuous.*

Problem 4.54 Prove that the process $f(t)$ in Theorems 4.51 and 4.52 is uniquely defined up to equivalency.

4.7 Change of measure and the Girsanov theorem

In this section, we assume again that $\mathcal{F}_t$ is a filtration such that an n-dimensional Wiener process $w(t)$ is $\mathcal{F}_t$ adapted and $w(t+\tau) - w(t)$ does not depend on $\mathcal{F}_t$. In particular, we allow that $\mathcal{F}_t$ is the filtration generated by the process $(w(t), \eta(t))$, where $\eta(t)$ is a random process that does not depend on $w(\cdot)$.

Let $\theta(t) = (\theta_1(t), \dots, \theta_n(t))$ be a bounded $\mathcal{F}_t$-adapted random process with values in $\mathbf{R}^n$ and with components from $\mathcal{L}_{22}$.

Let

$$\mathcal{Z} \triangleq \exp\left(-\int_0^T \theta(t)^\top dw(t) - \frac{1}{2}\int_0^T |\theta(t)|^2 dt \right). \tag{4.17}$$

Change of the probability measure

Proposition 4.55 *Let the process θ be bounded.*[4] *Then*

(i) $\mathbf{E}\mathcal{Z} = 1$.

4 Instead of boundedness of θ, we could assume that the less restrictive so-called Novikov's condition is satisfied:

$$\mathbf{E}\exp\left(\frac{1}{2}\int_0^T |\theta(t)|^2 dt\right) < +\infty.$$

Clearly, this condition is satisfied for all bounded processes θ.

(ii) Let mapping $\mathbf{P}_* : \mathcal{F} \to \mathbf{R}$ *be defined via the equation*

$$\frac{d\mathbf{P}_*}{d\mathbf{P}} = \mathcal{Z}. \tag{4.18}$$

Then $\mathbf{P}_*$ *is a probability measure on* $\mathcal{F}$ *equivalent to the original measure* $\mathbf{P}$.

Remember that (4.18) means that

$$\mathbf{P}_*(A) = \int_A \mathcal{Z}(\omega)\mathbf{P}(d\omega) \quad \forall A \in \mathcal{F},$$

i.e.,

$$\mathbf{E}_*\mathcal{Z} = \mathbf{E}\mathbb{I}_A\mathcal{Z} \quad \forall A \in \mathcal{F},$$

and

$$\mathbf{E}_*\xi = \mathbf{E}\mathcal{Z}\xi$$

for any integrable random variable ξ (see Section 1.4).

Proof of Proposition 4.55. Let us prove (i). We have that $\mathcal{Z} = y(T)$, where $y(t)$ is the solution of the equation

$$\begin{cases} dy(t) = -y(t)\theta(t)^\top dw(t), \\ y(0) = 1. \end{cases}$$

Then

$$y(T) = 1 + \int_0^T y(t)\theta(t)^\top dw(t).$$

Hence $\mathbf{E}\mathcal{Z} = \mathbf{E}y(T) = 1$. To prove (ii), it suffices to verify that all probability axioms are satisfied. For instance, we have that $\mathbf{P}_*(\Omega) = \mathbf{E}\mathcal{Z}\mathbb{I}_{\{\omega\in\Omega\}} = \mathbf{E}\mathcal{Z} = 1$. □

Example 4.56 We have $\mathbf{E}_*\mathcal{Z} = \mathbf{E}\mathcal{Z}\cdot\mathcal{Z} = \mathbf{E}\mathcal{Z}^2$.

Example 4.57 Let $n = 1$. We have that $\mathbf{E}_*w(T) = \mathbf{E}\mathcal{Z}w(T)$. Let θ be non-random and constant, then

$$\begin{aligned} \mathbf{E}_*w(T) &= \mathbf{E}w(T)\exp\left(-\theta w(T) - \frac{T}{2}\theta^2\right) \\ &= \frac{1}{\sqrt{2\pi T}}\int_{-\infty}^{+\infty} e^{-\frac{x^2}{2T}} x \exp\left(-\theta x - \frac{T}{2}\theta^2\right)dx. \end{aligned}$$

Girsanov's theorem

Let

$$w_*(t) \stackrel{\Delta}{=} w(t) + \int_0^t \theta(s)ds.$$

The following is a special case of the celebrated Girsanov's theorem.

Theorem 4.58 *Let the assumptions of Proposition 4.55 be satisfied, and let the measure $\mathbf{P}_*$ be defined via the equation*

$$\frac{d\mathbf{P}_*}{d\mathbf{P}} = \mathcal{Z}.$$

Then $w_*(t)$ is a Wiener process under $\mathbf{P}_*$.

Proof of Theorem 4.58. We are going to prove only that

$$\mathbf{E}_* e^{\int_0^T f(t)dw_*(t)} = \mathbf{E} e^{\int_0^T f(t)dw(t)} \tag{4.19}$$

for all deterministic continuous functions $f(\cdot) : [0, T] \to \mathbf{R}^{1\times n}$. (In fact, it suffices; it follows that $w(\cdot)$ and $w_*(\cdot)$ have the same distributions as processes in $(\Omega, \mathcal{F}, \mathbf{P})$ and $(\Omega, \mathcal{F}, \mathbf{P}_*)$ respectively.)

Let

$$\psi(t) \stackrel{\Delta}{=} f(t) - \theta(t)^\top. \quad \square$$

Proposition 4.59

$$\mathbf{E}\exp\left(-\frac{1}{2}\int_0^T |\psi(t)|^2 dt + \int_0^T \psi(t)dw(t)\right) = 1.$$

Proof of Proposition 4.59. Let $y(t)$ be the solution of the equation

$$\begin{cases} dy(t) = y(t)\psi(t)dw(t), \\ y(0) = 1. \end{cases}$$

Then

$$y(T) = 1 + \int_0^T y(t)\psi(t)dw(t).$$

Hence $\mathbf{E}y(T) = 1$. On the other hand,

$$y(T) = \exp\left(-\frac{1}{2}\int_0^T |\psi(t)|^2 dt + \int_0^T \psi(t)dw(t)\right).$$

This completes the proof of Proposition 4.59. $\square$

Proposition 4.60 *$\mathbf{E}\exp\left(\int_0^T f(t)dw(t)\right) = \exp\left(\frac{1}{2}\int_0^T |f(t)|^2 dt\right)$ for all deterministic continuous functions $f(\cdot) : [0,T] \to \mathbf{R}^{1\times n}$.*

Proof. Apply Proposition 4.59 with $\theta(t) \equiv 0$. □

Let us complete the proof of Theorem 4.58. We have that

$$\begin{aligned}
&\mathbf{E}_* \exp\left(\int_0^T f(t)dw_*(t)\right)\\
&\quad = \mathbf{E}_* \exp\left(\int_0^T f(t)[\theta(t)dt + dw(t)]\right)\\
&\quad = \mathbf{E}\exp\left(-\int_0^T \theta(t)^\top dw(t) - \frac{1}{2}\int_0^T |\theta(t)|^2 dt + \int_0^T f(t)[\theta(t)dt + dw(t)]\right)\\
&\quad = \mathbf{E}\exp\left(\int_0^T \psi(t)dw(t) - \frac{1}{2}\int_0^T |\theta(t)|^2 dt + \int_0^T f(t)\theta(t)dt\right)\\
&\quad = \mathbf{E}\exp\left(\int_0^T \psi(t)dw(t) - \frac{1}{2}\int_0^T |\theta(t)|^2 dt + \int_0^T f(t)\theta(t)dt\right.\\
&\qquad \left. - \frac{1}{2}\int_0^T |f(t)|^2 dt + \frac{1}{2}\int_0^T |f(t)|^2 dt\right)\\
&\quad = \exp\left(\frac{1}{2}\int_0^T |f(t)|^2 dt\right)\mathbf{E}\exp\left(\int_0^T \psi(t)dw(t) - \frac{1}{2}\int_0^T |\psi(t)|^2 dt\right).
\end{aligned}$$

It follows from Propositions 4.59 and 4.60 that

$$\mathbf{E}_* \exp\left(\int_0^T f(t)dw_*(t)\right) = \exp\left(\frac{1}{2}\int_0^T |f(t)|^2 dt\right) = \mathbf{E}\exp\left(\int_0^T f(t)dw(t)\right).$$

This completes the proof of Theorem 4.58. □

Example 4.61 Let us reconsider Example 4.57. We have that $w(T) = w_*(T) - \int_0^T \theta(s)ds$. Hence

$$\mathbf{E}_* w(T) = \mathbf{E}_*\left[w_*(T) - \int_0^T \theta(s)ds\right] = -\mathbf{E}_* \int_0^T \theta(s)ds.$$

(Remember that w_* is a Wiener process with respect to $\mathbf{P}_*$, hence $\mathbf{E}_* w_*(T) = 0$.) One can verify that the integral in Example 4.57 has the value $-\mathbf{E}_* \int_0^T \theta(s)ds = -T\theta$ for the case of non-random and constant θ.

4.8 Problems

In these problems, all processes are one-dimensional.

Ito integral

Problem 4.62 Let $f(t) = 2\cdot\mathbb{I}_{\{t\in[0.1,0.6]\}}$, $g(t) = -3\cdot\mathbb{I}_{\{t\in[0.2,0.8]\}}$. Find explicitly $\int_0^1 f(t)dw(t)$, $\int_0^1 g(t)dw(t)$ (i.e., express these integrals as functions of $w(\cdot)$).

Problem 4.63 Let f and g be defined in the previous problem. Find $\mathbf{E}\left(\int_0^1 f(t)dw(t)\right)^2$, $\mathbf{E}\left(\int_0^1 g(t)dw(t)\right)^2$, $\mathbf{E}\int_0^1 f(t)dw(t)\int_0^1 g(t)dw(t)$.

Problem 4.64 Let $f(t)=4\cdot\mathbb{I}_{\{t\in[0.1,0.4]\}}-4\cdot\mathbb{I}_{\{t\in[0.3,0.7]\}}$. Find explicitly $\int_0^1 f(t)dw(t)$ (i.e., express the integrals as functions of $w(\cdot)$).

Problem 4.65 Let f be defined in the previous problem. Let $g(t) = \mathbb{I}_{\{t\in[0.5,1]\}}$. Find $\mathbf{E}\int_0^1 f(t)dw(t)\int_0^1 g(t)dw(t)$.

Problem 4.66 Let $f(t)=e^t$, $g(t)=w(t)$. Find $\mathbf{E}\int_0^1 f(t)dw(t)$, $\mathbf{E}\left(\int_0^1 f(t)dw(t)\right)^2$, $\mathbf{E}\left(\int_0^1 g(t)dw(t)\right)^2$, $\mathbf{E}\int_0^1 f(t)dw(t)\int_0^1 g(t)dw(t)$.

Problem 4.67 Let $f(t)=e^t$, $g(t)=e^{-2t}$. Let $\mathcal{F}_t$ be the filtration generated by $w(t)$. Find $\mathbf{E}\{\int_0^5 f(s)dw(s)\,|\,\mathcal{F}_t\}$, $\mathbf{E}\left\{\left(\int_0^5 f(s)dw(s)\right)^2\,\Big|\,\mathcal{F}_t\right\}$, $\mathbf{E}\left\{\left(\int_0^5 g(s)dw(s)\right)^2\,\Big|\,\mathcal{F}_t\right\}$, $\mathbf{E}\left\{\int_0^5 f(s)dw(s)\int_0^5 g(s)dw(s)\,\Big|\,\mathcal{F}_t\right\}$.

Ito differential and Ito formula

Problem 4.68 Let $y(t) = w(t)^2$. Prove that $dy(t) = 2w(t)dw(t) + dt$.

Problem 4.69 Let $y(t) = w(t)^3$. Find $dy(t)$. Find $\mathbf{E}y(t)$.

Problem 4.70 Let $y(t) = w(t)^4$. Find $dy(t)$. Find $\mathbf{E}y(t)$.

Problem 4.71 Let $y(t) = \sin w(t)$. Find $dy(t)$.

Problem 4.72 Let $y(t) = \sin z(t)$, where $z(t) = w(t)^2 + t$. Find $dy(t)$.

Problem 4.73 Let $y(t) = \cos z(t)$, where $z(t) = w(t)^2 - t$. Find $dy(t)$.

Problem 4.74 Let $y(t) = e^{w(t)}$. Prove that $dy(t) = y(t)dw(t) + (1/2)y(t)dt$. Find $dy(t)$. Find $\mathbf{E}y(t)$.

Problem 4.75 Let $y(t) = e^{w(t)-t/2}$. Prove that $dy(t) = ydw(t)$. Find $\mathbf{E}y(t)$.

Problem 4.76 Let $y(t) = e^{2w(t)+3t}$. Find $dy(t)$. Find $\mathbf{E}y(t)$.

Problem 4.77 Let $y(t) = e^{w(t)}$, $x(t) = w(t)$. Find $d(y(t)x(t))$.

Ito equations

Solve Problems 4.38–4.41.

Problem 4.78 For Problem 4.41, find equations for $\mathbf{E}y(t)$ and $\mathbf{E}y(t)^2$.

Problem 4.79 Show that the equation

$$\begin{cases} dy(t) = dt/2 + \sqrt{2y(t)}dw(t) \\ y(0) = 0 \end{cases} \tag{4.20}$$

has a solution $y(t) = w(t)^2/2$, $t \geq 0$. Can we apply the existence and uniqueness theorem 4.36 to verify the uniqueness?

Markov processes and Kolmogorov equations

Problem 4.80 Find the probabilistic representation of solutions of the Cauchy problem for the parabolic equation

$$\frac{du}{dt}(x,t) + \frac{d^2u}{dx^2}(x,t) = 0, \quad x \in \mathbf{R},\ t < T,$$
$$u(x,T) = \Psi(x).$$

Problem 4.81 Let $y(t) = y^{a,s}(t)$ be a solution of the stochastic differential equation

$$\begin{cases} dy(t) = 0.7y(t)dw(t), & t > s, \\ y(s) = a. \end{cases} \tag{4.21}$$

Find deterministic equations for the function $u(x,s) = \mathbf{E}\Psi(y^{x,s}(T))$ for a given function Ψ.

Problem 4.82 Let $y(t) = y^{a,s}(t)$ be a solution of the stochastic differential equation

$$\begin{cases} dy(t) = \sin y(t)dt + \cos(y(t) - 1)dw(t), & t > s, \\ y(s) = a. \end{cases} \tag{4.22}$$

Find deterministic equations for the function $u(x,s) = \mathbf{E}\Psi(y^{x,s}(T))$ for a given function Ψ.

Problem 4.83 Let $f(x,t) \equiv \sin x$, $b(x,t) \equiv \sin(x-1)$. Find deterministic equations for the function $u(x,s) = \mathbf{E}\Psi(y^{x,s}(T))$ for a function Ψ.

Problem 4.84 Given a function Ψ, find the probabilistic representation of the solution of the Cauchy problem for the parabolic equation

$$\frac{\partial u}{\partial t}(x,t) + \frac{\partial^2 u}{\partial x^2}(x,t) = 0, \quad x \in \mathbf{R},\ t < T,$$

$$u(x,T) = \Psi(x).$$

5 Continuous time market models

In this chapter, the most mainstream models of markets with continuous time are studied. These models are based on the theory of stochastic integrals (stochastic calculus); stock prices are represented via stochastic integrals. Core concepts and results of mathematical finance are given (including self-financing strategies, replicating, arbitrage, risk-neutral measures, market completeness, and option price).

5.1 Continuous time model for stock price

We assume that we are given a standard complete probability space $(\Omega, \mathcal{F}, \mathbf{P})$ (see Chapter 1). Sometimes we shall address $\mathbf{P}$ as the *original probability measure*. Other measures will also be used.

Consider a risky asset (stock, bond, foreign currency unit, etc.) with time-series prices $S_1, S_2, S_3, \ldots$, for example daily prices. The premier model of price evolution is such that $S_k = S(t_k)$, where $S(t)$ is a continuous time Ito process. (Note that Ito processes are pathwise continuous. For a more general model, continuous time process $S(t)$ may have jumps; this case will not be considered here.)

We consider evolution of the price $S(t)$ for $t \in [0, T]$, where t is time, T is some terminal time.

The initial price $S_0 > 0$ is a given non-random value, and the evolution of $S(t)$ is described by the following *Ito equation*:

$$dS(t) = S(t)(a(t)dt + \sigma(t)dw(t)). \tag{5.1}$$

Here $w(t)$ is a (one-dimensional) Wiener process, and a and σ are market parameters.

Sometimes in the literature $S(t)$ is called a geometric Brownian motion (for the case of non-random and constant a, σ), sometimes $\ln S(t)$ is also said to be a Brownian motion. Mathematicians prefer to use the term 'Brownian motion' for $w(t)$ only (i.e., Brownian motion is the same as a Wiener process).

Definition 5.1 *In (5.1), $a(t)$ is said to be the appreciation rate, $\sigma(t)$ is said to be the volatility.*

Note that, in terms of more general stochastic differential equations, the coefficient for dt (i.e., $a(t)S(t)$) is said to be the drift (or the drift coefficient), and the coefficient for $dw(t)$ (i.e., $\sigma(t)S(t)$) is said to be the diffusion coefficient.

Definition 5.2 *If $\sigma(t)$ is such that $\sigma(t) \neq 0$ a.e. (i.e., for almost every t with probability 1), then equation (5.1) (and the market model) is said to be non-degenerate.*

We assume that there exists a random process $\eta(t)$ that does not depend on $w(\cdot)$. This process describes additional random factors presented in the model besides the driving Wiener process $w(t)$.

Let $\mathcal{F}_t$ be the filtration generated by $(w(t), \eta(t))$, and let $\mathcal{F}_t^w$ be the filtration generated by the process $w(t)$ only.

It follows that $\mathcal{F}_t^w \subseteq \mathcal{F}_t$ and that $w(t+\tau) - w(t)$ does not depend on $\mathcal{F}_t$ for all t and $\tau > 0$. (Note that the case when $\mathcal{F}_t \equiv \mathcal{F}_t^w$ is not excluded.)

We assume that the process $(a(t), \sigma(t))$ is $\mathcal{F}_t$-adapted. In particular, it follows that $(a(t), \sigma(t))$ does not depend on $w(t+\tau) - w(t)$ for all t and $\tau > 0$.

Without loss of generality, we assume that $\mathcal{F} = \mathcal{F}_T$.

Remark 5.3 The assumptions imposed imply that the vector $(r(t), a(t), \sigma(t))$ can be presented as a deterministic function of $(w(s), \eta(s))|_{s\in[0,t]}$.

Let us discuss some basic properties of the Ito equation (5.1).

Lemma 5.4

$$S(t) = S(0)\exp\left(\int_0^t a(s)ds - \frac{1}{2}\int_0^t \sigma(s)^2 ds + \int_0^t \sigma(s)dw(s)\right).$$

Proof follows from the Ito formula (see Problem 4.40). □

Note that the stochastic integral above is well defined.

The process $S(t)$ has the following properties (for the case of non-zero σ):

- sample paths maintain continuity;
- sample paths are non-differentiable;
- if a, σ are deterministic and constant, then

$$\mathrm{Var}\,\frac{S(t+\Delta t)}{S(t)} = e^{\sigma^2 \cdot \Delta t} \quad \forall t > 0,\ \Delta t > 0;$$

- if a, σ are deterministic, then the probability distribution of $S(t)$ is log-normal (i.e., its logarithm follows a normal law);
- if a, σ are deterministic, then the relative increments $[S(t) - S(\tau)]/S(\tau)$ are independent from the past prices $S(s)|_{s\le\tau}$ for $0 \le \tau < t$;

- if a, σ are deterministic and constant, then the probability distribution of relative increments does not depend on time shift. More precisely, the probability distribution of $[S(t) - S(\tau)]/S(\tau)$ is identical to the distribution of $[S(t-\tau) - S(0)]/S(0)$, $0 \le \tau < t$.

5.2 Continuous time bond–stock market model

The case of the market with a non-zero interest rate for borrowing can be described via the following bond–stock model.

We introduce a market model consisting of the risk-free bond or bank account with price $B(t)$ and the risky stock with the price $S(t)$, $t > 0$. The initial prices $S(0) > 0$ and $B(0) > 0$ are given non-random variables. We assume that the bond price is

$$B(t) = B(0)\exp\left(\int_0^t r(s)ds\right), \tag{5.2}$$

where $r(t)$ is the process of the risk-free interest rate. We assume that the process $r(t)$ is $\mathcal{F}_t$-adapted (in particular, it follows that $r(t)$ does not depend on $w(t+\tau) - w(t)$ for all t, $\tau > 0$). Typically, it suffices to consider non-negative processes $r(t)$ (however, we do not assume this, because it can be restrictive for some models, especially for bond markets).

Let $X(0) > 0$ be the initial wealth at time $t = 0$, and let $X(t)$ be the wealth at time $t > 0$. We assume that, for $t \ge 0$,

$$X(t) = \beta(t)B(t) + \gamma(t)S(t). \tag{5.3}$$

Here $\beta(t)$ is the quantity of the bond portfolio, $\gamma(t)$ is the quantity of the stock portfolio. The pair $(\beta(\cdot), \gamma(\cdot))$ describes the state of the bond–stocks portfolio at time t. Each of these pairs is called a strategy (portfolio strategy).

We consider the problem of trading or choosing a strategy in a class of strategies that does not use future values of $(S(t), r(t))$. Some constraints will be imposed on current operations in the market, or in other words, on strategies.

Definition 5.5 *A pair $(\beta(\cdot), \gamma(\cdot))$ is said to be an admissible strategy if $\beta(t)$ and $\gamma(t)$ are random processes adapted to the filtration $\mathcal{F}_t$ and such that*

$$\mathbf{E}\int_0^T \left(B(t)^2\beta(t)^2 + S(t)^2\gamma(t)^2\right) dt < +\infty. \tag{5.4}$$

Definition 5.6 *A pair $(\beta(\cdot), \gamma(\cdot))$ is said to be a self-financing strategy, if*

$$dX(t) = \beta(t)dB(t) + \gamma(t)dS(t). \tag{5.5}$$

Note that condition (5.4) ensures that the process $X(t)$ is well defined by equation (5.5) as an Ito process.

We allow negative $\beta(t)$ and $\gamma(t)$, meaning borrowing and short positions.

We shall consider admissible self-financing strategies only.

Remark 5.7 In literature, a definition of admissible strategies may include requirements that the risk is bounded. An example of this requirement is the following: there exists a constant C such that $X(t) \geq C$ for all t a.s. For simplicity, we do not require this.

Remark 5.8 The case of $r(t) \equiv 0$ corresponds to the market model with free borrowing.

Some strategies

Example 5.9 For risk-free, 'keep-only-bonds', the strategy is such that the portfolio contains only the bonds, $\gamma(t) \equiv 0$, and the corresponding total wealth is $X(t) \equiv \beta_0 B(t) \equiv \exp(\int_0^t r(s)ds)X(0)$.

Example 5.10 *Buy-and-hold strategy* is a strategy when $\gamma(t) > 0$ does not depend on time. This strategy ensures a gain when the stock price is increasing.

Example 5.11 *Merton's type strategy* is a strategy in a closed-loop form when $\gamma(t) = \mu(t)\theta(t)X(t)$, where $\mu(t) > 0$ is a coefficient, $X(t)$ is the wealth, $\theta(t) \triangleq \sigma(t)^{-1}[a(t) - r(t)]$ is the so-called market price of the risk process. This strategy is important since it is optimal for certain optimal investment problems (including maximization of $\mathbf{E} \ln X(T)$).

Example 5.12 *('buy low, sell high' rule).* Let $T = +\infty$. Consider a strategy when $\gamma(t) = \mathbb{I}_{t\in[0,\tau]}$, where $C > 0$ is a constant, $\tau = \min\{t > 0 : S(t) = K\}$. Here $K > S(0)$ is a 'high' price $K > S(0)$, or the goal price. Let $X(0) = S(0)$ and $\beta(t) \equiv 0$, then $X(\tau) = S(\tau) = K > X(0)$. This strategy has interesting mathematical features for the non-degenerate diffusion model (i.e., when $\sigma(t) \geq c > 0$ for some constant $c > 0$). In this case, we have that $\mathbf{P}(\tau < +\infty) = 1$ for all $K > 0$ (i.e., any 'high' price will be achieved with probability 1. However, $\mathbf{E}\tau = +\infty$ (i.e., the risk-free gain is achieved for stopping time that is not reasonably small). (See related Problem 5.83 below.)

5.3 The discounted wealth and stock prices

For the trivial, risk-free, 'keep-only-bonds' strategy, the portfolio contains only the bonds, $\gamma(t) \equiv 0$, and the corresponding total wealth is $X(t) \equiv \beta_0 B(t) \equiv \exp(\int_0^t r(s)ds)X(0)$. Some loss is possible for a strategy that deals with risky assets. It is natural to estimate the loss and gain by comparing it with the results for the 'keep-only-bonds' strategy.

Definition 5.13 *The process* $\tilde{X}(t) \triangleq \exp(-\int_0^t r(s)ds)X(t)$, $\tilde{X}(0) = X(0)$, *is called the discounted wealth (or the normalized wealth).*

Definition 5.14 *The process* $\tilde{S}(t) \triangleq \exp(-\int_0^t r(s)ds)S(t)$, $\tilde{S}_0 = S_0$, *is called the discounted stock price (or the normalized stock price).*

Let $\tilde{a}(t) \triangleq a(t) - r(t)$.

Proposition 5.15 $d\tilde{S}(t) = \tilde{S}(t)(\tilde{a}(t)dt + \sigma(t)dw(t))$.

The proof is straightforward.

Theorem 5.16 *The property of self-financing (5.5) is equivalent to*

$$d\tilde{X}(t) = \gamma(t)d\tilde{S}(t), \tag{5.6}$$

i.e.,

$$\tilde{X}(t) = X(0) + \int_0^t \gamma(s)d\tilde{S}(s). \tag{5.7}$$

Proof of Theorem 5.16. Let $(\tilde{X}(t), \gamma(t))$ be a process such that (5.6) holds. Then it suffices to prove that $X(t) \triangleq \exp(\int_0^t r(s)ds)\tilde{X}(t)$ is the wealth corresponding to the self-financing strategy $(\beta(\cdot), \gamma(\cdot))$, where $\beta(t) = (X(t) - \gamma(t)X(t))B(t)^{-1}$.

We have that

$$\begin{aligned}
dX(t) &= \exp\left(\int_0^t r(s)ds\right)d\tilde{X}(t) + r(t)X(t)dt \\
&= \exp\left(\int_0^t r(s)ds\right)\gamma(t)d\tilde{S}(t) + r(t)X(t)dt \\
&= \exp\left(\int_0^t r(s)ds\right)\gamma(t)\tilde{S}(t)[\tilde{a}(t)dt + \sigma(t)dw(t)] + r(t)X(t)dt \\
&= \gamma(t)S(t)[\tilde{a}(t)dt + \sigma(t)dw(t)] + r(t)[\gamma(t)S(t) + \beta(t)B(t)]dt \\
&= \gamma(t)S(t)[a(t)dt + \sigma(t)dw(t)] + r(t)\beta(t)B(t)dt \\
&= \gamma(t)dS(t) + \beta(t)dB(t).
\end{aligned}$$

This completes the proof. □

Thanks to Theorem 5.16, we can reduce many problems for markets with non-zero interest for borrowing to the simpler case of the market with zero interest rate (free borrowing). In particular, it makes calculation of the wealth for a given strategy easier.

Example 5.17 Let $r(t) \equiv r$. Let $\gamma(t) = \mathbb{I}_{\{t\in[t_1,t_2]\}}$, where $0 \le t_1 < t_2 \le T$. Then

$$\tilde{X}(T) = X(0) + \int_0^T \gamma(t)d\tilde{S}(t) = X(0) + \int_{t_1}^{t_2} d\tilde{S}(t) = X(0) + \tilde{S}(t_2) - \tilde{S}(t_1),$$

and $X(T) = e^{rT}\tilde{X}(T)$.

Example 5.18 Let $r(t) \equiv r$, $a(t) \equiv a$, $\sigma(t) \equiv \sigma$ be constant. Let $\gamma(t) = k\tilde{S}(t)^{-1}$, where $k \in \mathbf{R}$. Then

$$\begin{aligned}\tilde{X}(T) &= X(0) + \int_0^T \gamma(t)d\tilde{S}(t)\\ &= X(0) + \int_0^T k\tilde{S}(t)^{-1}d\tilde{S}(t) = X(0) + \int_0^T k(\tilde{a}dt + \sigma dw(t))\\ &= X(0) + k\tilde{a}T + k\sigma \int_0^T dw(t) = X(0) + k\tilde{a}T + k\sigma w(T).\end{aligned}$$

Clearly, $X(T) = e^{rT}\tilde{X}(T)$. It can be seen that the random variable $\tilde{X}(T) - X(0)$ is Gaussian with the law $N(k\tilde{a}T, k^2\sigma^2 T)$.

Example 5.19 Let $r(t) \equiv r$, $a(t) \equiv a$, $\sigma(t) \equiv \sigma$ be constant. Let $\gamma(t) = k\tilde{a}\tilde{S}(t)^{-1}$, where $k > 0$, $\tilde{a} = a - r$. Then

$$\begin{aligned}\tilde{X}(T) &= X(0) + \int_0^T \gamma(t)d\tilde{S}(t) = X(0) + \int_0^T k\tilde{a}\tilde{S}(t)^{-1}d\tilde{S}(t)\\ &= X(0) + k\int_0^T \tilde{a}(\tilde{a}dt + \sigma dw(t)) = X(0) + k\tilde{a}^2T + k\tilde{a}\sigma\int_0^T dw(t)\\ &= X(0) + k\tilde{a}^2T + k\tilde{a}\sigma w(T).\end{aligned}$$

Clearly, $X(T) = e^{rT}\tilde{X}(T)$. It can be seen that the random variable $\tilde{X}(T) - X(0)$ is Gaussian with the law $N(k\tilde{a}^2T, k^2a^2\sigma^2T)$. It can also be seen that this strategy gives positive average gain if $\tilde{a} \neq 0$.

For simplicity, one can assume for the first reading that (a, σ) is non-random and constant, $\eta(t) \equiv 0$, and that $r(t) \equiv 0$, $B(t) \equiv B(0)$, $\tilde{X}(t) \equiv X(t)$, and $\tilde{S}(t) \equiv S(t)$, everywhere in this chapter. After that, one can read this chapter again taking into account the general case.

5.4 Risk-neutral measure

Definitions

Remember that $\mathcal{F} = \mathcal{F}_T$, and $\mathcal{F}_t$ is the filtration generated by the process $(w(t), \eta(t))$, where $\eta(\cdot)$ is a process independent from $w(\cdot)$ that describes additional random factors presented in the model besides the driving Wiener process (see also Remark 5.3).

Definition 5.20 *Let $\mathbf{P}_* : \mathcal{F} \to [0,1]$ be a probability measure such that the process $\tilde{S}(t)$ is a martingale with respect to the filtration $\mathcal{F}_t$ for $\mathbf{P}_*$. Then $\mathbf{P}_*$ is said to be a risk-neutral probability measure for the bond–stock market (5.1), (5.2).*

In the literature, a risk-neutral measure is also called a martingale measure.
As usual, $\mathbf{E}_*$ denotes the corresponding expectation.
In particular, $\mathbf{E}_*\{\tilde{S}_\tau \mid \tilde{S}(\cdot)|_{[0,t]}\} = \tilde{S}(t)$ for all $\tau > t$.

Definition 5.21 *If a risk-neutral probability measure $\mathbf{P}_*$ is equivalent to the original measure $\mathbf{P}$, we call it an equivalent risk-neutral measure.*

Market price of risk

Remember that $\tilde{a}(t) \triangleq a(t) - r(t)$. Let a process θ be a solution of the equation

$$\sigma(t)\theta(t) = \tilde{a}(t). \tag{5.8}$$

This process θ is called the *market price of risk* process; this term came from optimal portfolio selection theory. If the market is non-degenerate, i.e., $\sigma(t) \neq 0$, then

$$\theta(t) = \sigma(t)^{-1}\tilde{a}(t) = \sigma(t)^{-1}[a(t) - r(t)] \quad \text{a.e.}$$

Up to the end of this chapter, we assume that the following condition is satisfied.

Condition 5.22 The market price of risk process exists for a.e. t, ω, and there exists a constant $c > 0$ such that $|\theta(t,\omega)| \leq c$ a.s. for a.e. t.[1]

Clearly, this condition ensures that if $\sigma(t) = 0$ then $\tilde{a}(t) = 0$ a.e., i.e., $a(t) = r(t)$ for a.e. t a.s.

1 Instead of Condition 5.22, we could assume that the less restrictive Novikov's condition is satisfied. We say that Novikov's condition is satisfied if

$$\mathbf{E}\exp\left(\frac{1}{2}\int_0^T |\theta(t)|^2 dt\right) < +\infty.$$

Clearly, Novikov's condition is satisfied if Condition 5.22 is satisfied.

It follows that if the market is non-degenerate (i.e., $|\sigma(t)| \geq \text{const.} > 0$), and the process $(r(t), \sigma(t), a(t))$ is bounded, then Condition 5.22 is satisfied.

A measure $\mathbf{P}_*$ defined by the market price of risk

Let

$$\mathcal{Z} \triangleq \exp\left(-\int_0^T \theta(t)dw(t) - \frac{1}{2}\int_0^T |\theta(t)|^2 dt\right). \tag{5.9}$$

By Proposition 4.55, it follows that

(i) $\mathbf{E}\mathcal{Z} = 1$;
(ii) If the mapping $\mathbf{P}_* : \mathcal{F} \to \mathbf{R}$ is defined via the equation

$$\frac{d\mathbf{P}_*}{d\mathbf{P}} = \mathcal{Z}, \tag{5.10}$$

then $\mathbf{P}_*$ is a probability measure on $\mathcal{F}$ equivalent to the original measure $\mathbf{P}$.

Remember that (5.10) means that $\mathbf{P}_*(A) = \int_A \mathcal{Z}(\omega)\mathbf{P}(d\omega)$ for any $A \in \mathcal{F}$, and $\mathbf{E}_*\xi = \mathbf{E}\mathcal{Z}\xi$ for any integrable random variable ξ.

Application of the Girsanov theorem

Let

$$w_*(t) \triangleq w(t) + \int_0^t \theta(s)ds.$$

Here θ is defined by (5.8).

Note that

$$\tilde{a}(t)dt + \sigma(t)dw(t) = \sigma(t)dw_*(t).$$

Hence

$$d\tilde{S}(t) = \tilde{S}(t)[\tilde{a}(t)dt + \sigma(t)dw(t)] = \tilde{S}(t)\sigma(t)dw_*(t).$$

$\mathbf{P}_*$ *as an equivalent risk-neutral measure*

Theorem 5.23 *Let Condition 5.22 be satisfied, and let the measure $\mathbf{P}_*$ be defined by equation (5.10). Then*

(i) $w_*(t)$ is a Wiener process under $\mathbf{P}_*$;
(ii) $\mathbf{P}_*$ is an equivalent risk-neutral measure.

Proof. Statement (i) follows from the Girsanov theorem (4.58), as well as the statement that $\mathbf{P}_*$ is equivalent to $\mathbf{P}$. Let us prove the rest of part (ii).

We have that

$$d\tilde{S}(t) = \tilde{S}(t)[\tilde{a}(t)dt + \sigma(t)dw(t)] = \tilde{S}(t)\sigma(t)dw_*(t).$$

By Theorem 4.58, $w_*(t)$ is a Wiener process under $\mathbf{P}_*$. Then

$$\begin{aligned}
\mathbf{E}_*\{\tilde{S}(t)\,|\,\mathcal{F}_s\} &= \mathbf{E}_*\left\{S_0 + \int_0^t \tilde{S}(\tau)\sigma(\tau)dw_*(\tau)\,\Big|\,\mathcal{F}_s\right\} \\
&= S(0) + \mathbf{E}_*\left\{\int_0^t \tilde{S}(\tau)\sigma(\tau)dw_*(\tau)\,\Big|\,\mathcal{F}_s\right\} \\
&= S_0 + \int_0^s \tilde{S}(\tau)\sigma(\tau)dw_*(\tau) \\
&= \tilde{S}(s).
\end{aligned}$$

This completes the proof. □

Theorem 5.24 *Let Condition 5.22 be satisfied, and let $\mathbf{P}_*$ be the equivalent risk-neutral measure defined in Theorem 4.58. For any admissible self-financing strategy, the corresponding discounted wealth $\tilde{X}(t)$ is a martingale with respect to $\mathcal{F}_t$ under $\mathbf{P}_*$.*

Proof. We have that

$$\begin{aligned}
d\tilde{S}(t) &= \tilde{S}(t)\sigma(t)dw_*(t), \\
d\tilde{X}(t) &= \gamma(t)d\tilde{S}(t) = \gamma(t)\tilde{S}(t)\sigma(t)dw_*(t),
\end{aligned}$$

where $\gamma(t)$ is the number of shares. By Girsanov's theorem, $w_*(t)$ is a Wiener process under $\mathbf{P}_*$. Then

$$\begin{aligned}
\mathbf{E}_*\{\tilde{X}(t)\,|\,\mathcal{F}_s\} &= \mathbf{E}_*\left\{X_0 + \int_0^t \gamma(\tau)\tilde{S}(\tau)\sigma(\tau)dw_*(\tau)\,\Big|\,\mathcal{F}_s\right\} \\
&= X_0 + \mathbf{E}_*\left\{\int_0^t \gamma(\tau)\tilde{S}(\tau)\sigma(\tau)dw_*(\tau)\,\Big|\,\mathcal{F}_s\right\} \\
&= X_0 + \int_0^s \gamma(\tau)\tilde{S}(\tau)\sigma(\tau)dw_*(\tau) \\
&= \tilde{X}(s).
\end{aligned}$$

This completes the proof. □

5.5 Replicating strategies

Remember that $T > 0$ is given.

Let ψ be a random variable.

Definition 5.25 *Let the initial wealth $X(0)$ be given, and let a self-financing strategy $(\beta(\cdot), \gamma(\cdot))$ be such that $X(T) = \psi$ a.s. for the corresponding wealth. Then the claim ψ is called replicable (attainable, redundant), and the strategy is said to be a replicating strategy (with respect to this claim).*

Definition 5.26 *Let the initial wealth $X(0)$ be given, and let a self-financing strategy $(\beta(\cdot), \gamma(\cdot))$ be such that $X(T) \geq \psi$ a.s. for the corresponding wealth. Then the strategy is said to be a super-replicating strategy.*

Theorem 5.27 *Let Condition 5.22 be satisfied, and let $\mathbf{P}_*$ be the equivalent risk-neutral measure such as defined in Theorem 5.23. Let ψ be an $\mathcal{F}_T$-measurable random variable such that $\mathbf{E}_*\psi^2 < +\infty$. Let the initial wealth $X(0)$ and a self-financing strategy $(\beta(\cdot), \gamma(\cdot))$ be such that $X(T) = \psi$ a.s. for the corresponding wealth. Then*

$$X(0) = \mathbf{E}_* \exp\left(-\int_0^T r(s)ds\right)\psi.$$

Proof. Clearly, $X(T) = \psi$ iff $\tilde{X}(T) = \exp(-\int_0^T r(s)ds)\psi$ a.s. We have that

$$\begin{aligned}
\mathbf{E}_*\tilde{X}(T) = \mathbf{E}_* \exp\left(-\int_0^T r(s)ds\right)\psi &= \mathbf{E}_*\left(X(0) + \int_0^T \gamma(t)d\tilde{S}(t)\right) \\
&= X(0) + \mathbf{E}_* \int_0^T \gamma(t)d\tilde{S}(t) \\
&= X(0) + \mathbf{E}_* \int_0^T \gamma(t)\tilde{S}(t)\sigma(t)dw_*(t) \\
&= X(0).
\end{aligned}$$

We have used here the fact that $w_*(t)$ is a Wiener process under $\mathbf{P}_*$, and $\int \cdot dw_*$ is an Ito integral under $\mathbf{P}_*$, so $\mathbf{E}_* \int \cdot dw_* = 0$. □

First application: the uniqueness of the replicating strategy

Theorem 5.28 *Let Condition 5.22 be satisfied, and let $\mathbf{P}_*$ be the equivalent risk-neutral measure defined in Theorem 5.23. Let ψ be an $\mathcal{F}_T$-measurable random variable, $\mathbf{E}_*\psi^2 < +\infty$. Let the initial wealth $X(0)$ and a self-financing strategy $(\beta(\cdot), \gamma(\cdot))$ be such that $X(T) = \psi$ a.s. for the corresponding wealth $X(t)$. Then the initial wealth $X(0)$ is uniquely defined. Moreover, the processes $X(t)$ and $\sigma(t)\gamma(t)$*

are uniquely defined up to equivalency. If $\sigma(t) \neq 0$ for a.e. t, then the replicating strategy and the corresponding wealth process $X(t)$ are uniquely defined up to equivalency.

Proof. Let the initial wealth $X^{(i)}(0)$ and the strategy $(\beta^{(i)}(\cdot), \gamma^{(i)}(\cdot))$ be such that $X^{(i)}(T) = \psi$ a.s. for the corresponding wealth $X^{(i)}(t)$, $i = 1, 2$.

Let $\tilde{X}^{(i)}(t)$ be the corresponding discounted wealth.

Set

$$\Gamma(t) \triangleq \gamma^{(1)}(t) - \gamma^{(2)}(t), \quad Y(t) \triangleq \tilde{X}^{(1)}(t) - \tilde{X}^{(2)}(t).$$

We have that $Y(T) = 0$ a.s. Hence

$$Y(T) = Y(0) + \int_0^T \Gamma(t) d\tilde{S}(t) = 0.$$

Then $Y(t) = \mathbf{E}_*\{Y(T) \mid \mathcal{F}_t\} = 0$ a.s., $Y(0) = 0$, and

$$\int_0^T \Gamma(t) d\tilde{S}(t) = 0.$$

Hence

$$0 = \mathbf{E}_* \left[\int_0^T \Gamma(t) d\tilde{S}(t) \right]^2 = \mathbf{E}_* \left[\int_0^T \Gamma(t)\tilde{S}(t)\sigma(t) dw_*(t) \right]^2$$
$$= \mathbf{E}_* \int_0^T |\Gamma(t)\tilde{S}(t)\sigma(t)|^2 dt = \mathbf{E}_* \int_0^T \Gamma(t)^2 \cdot |\tilde{S}(t)\sigma(t)|^2 dt.$$

Hence $\Gamma(t)\sigma(t) = 0$ for a.e. t a.s. If $\sigma(t) \neq 0$ a.e., then $|\tilde{S}(t)\sigma(t)|^2 \neq 0$, and $(\beta^{(1)}(t), \gamma^{(1)}(t)) = (\beta^{(2)}(t), \gamma^{(2)}(t))$ for a.e. t a.s. □

5.6 Arbitrage possibilities and arbitrage-free markets

Similarly to the case of the discrete time market, we define arbitrage as a possibility of a risk-free positive gain. The formal definition is as follows.

Definition 5.29 *Let $T > 0$ be given. Let $(\beta(\cdot), \gamma(\cdot))$ be an admissible self-financing strategy, and let $\tilde{X}(t)$ be the corresponding discounted wealth. If*

$$\mathbf{P}(\tilde{X}(T) \geq X(0)) = 1, \quad \mathbf{P}(\tilde{X}(T) > X(0)) > 0, \tag{5.11}$$

then this strategy is said to be an arbitrage strategy. If there exists an arbitrage strategy, then we say that the market model allows an arbitrage.

As we have mentioned in Chapter 3, we are interested in models without arbitrage possibilities. If a model allows arbitrage, then it is usually not useful (despite the fact that arbitrage opportunities could exist occasionally in real-life market situations).

Problem 5.30 Let there exist t_1 and t_2 such that $0 \le t_1 < t_2 \le T$ and $\sigma(t) = 0$, $\tilde{a}(t) \neq 0$ for $t \in (t_1, t_2)$ a.s. Prove that this market model allows arbitrage. *Hint:* take $\gamma(t) = \tilde{a}(t)\mathbb{I}_{t\in(t_1,t_2)}$.

Theorem 5.31 *Let a market model be such that Condition 5.22 is satisfied (in particular, this means that there exists an equivalent risk-neutral probability measure that is equivalent to the original measure* **P***). Then the market model does not allow arbitrage.*

Proof. Let $(\beta(\cdot), \gamma(\cdot))$ be a self-financing admissible strategy that ensures arbitrage, i.e., it is such that (5.11) holds for the corresponding discounted wealth. Let $\mathbf{P}_*$ be the equivalent risk-neutral measure defined in Theorem 4.58. Then

$$\mathbf{P}_*(\tilde{X}(T) \ge X(0)) = 1, \quad \mathbf{P}_*(\tilde{X}(T) > X(0)) > 0. \tag{5.12}$$

Hence

$$\mathbf{E}_*\tilde{X}(T) > X(0). \tag{5.13}$$

But

$$\begin{aligned}\mathbf{E}_*\tilde{X}(T) &= X(0) + \mathbf{E}_* \int_0^T \gamma(t)d\tilde{S}(t) \\ &= X(0) + \mathbf{E}_* \int_0^T \gamma(t)\tilde{S}(t)\sigma(t)dw_*(t) = X(0),\end{aligned}$$

since w_* is a Wiener process under $\mathbf{P}_*$. This contradicts (5.13). We have used again the fact that $w_*(t)$ is a Wiener process under $\mathbf{P}_*$, and $\int \cdot dw_*$ is an Ito integral under $\mathbf{P}_*$, i.e., $\mathbf{E}_* \int \cdot dw_* = 0$. □

Problem 5.32 Prove that an equivalent probability measure does not exist for Problem 5.30. (Suggest a proof that is not based on Theorem 5.31.) Assume that $(\tilde{a}(t), \sigma(t))$ is a non-random continuous function. *Hint:* use the fact that there exists $\varepsilon > 0$ such that the sign of $a(t)$ is constant for $t \in (t_1, t_1 + \varepsilon)$.

Remark 5.33 We can repeat here Remark 3.32 regarding the *'fundamental theorem of asset pricing'*.

5.7 A case of complete market

Let $\mathcal{F}_t^{S,r}$ be the filtration generated by the process $(S(t), r(t))$. (Note that $\mathcal{F}_t^{S,r} \subseteq \mathcal{F}_t$, and, for the general case, $\mathcal{F}_t$ is larger than $\mathcal{F}_t^{S,r}$).

In fact, any $\mathcal{F}_T^{S,r}$-measurable random variable ψ can be presented as $\psi = F(S(\cdot), B(\cdot))$ for a certain mapping $F(\cdot) : C(0,T) \times C(0,T) \to \mathbf{R}$ (see related Theorem 1.45).

Definition 5.34 *A market model is said to be complete if any $\mathcal{F}_T^{S,r}$-measurable random claim ψ such that $\mathbf{E}_*|\psi|^2 < +\infty$ for some risk-neutral measure $\mathbf{P}_*$ is replicable with some initial wealth.*

Theorem 5.35 *If a market model is complete and there exists an equivalent risk-neutral measure, then this measure is unique (as a measure of $\mathcal{F}_T^{S,r}$).*

Proof. Let $A \in \mathcal{F}_T^{S,r}$. By the assumption, the claim $\exp(\int_0^T r(t)dt)\mathbb{I}_A$ is replicable with some initial wealth $X_A(0)$ ($\mathbb{I}_A$ is the indicator function of A). By Theorem 5.28, this $X_A(0)$ is uniquely defined. By Theorem 5.27, $\mathbf{E}_*\mathbb{I}_A = X_A(0)$ for any risk-neutral measure $\mathbf{P}_*$. Therefore, $\mathbf{P}_*$ is uniquely defined on $\mathcal{F}_T^{S,r}$. □

5.8 Completeness of the Black–Scholes model

The so-called Black–Scholes model (Black and Scholes 1973) is such that the vector $(r(t), a(t), \sigma(t)) \equiv (r, a, \sigma)$ is non-random and constant, $\sigma \neq 0$. For this model, we assume that the filtration $\mathcal{F}_t$ is generated by $w(t)$ (or by $S(t)$, or by $\tilde{S}(t)$).

Let w_* and $\mathbf{P}_*$ be defined as above, i.e., $d\mathbf{P}_*/d\mathbf{P} = \mathcal{Z}$. Remember that $w_*(t)$ is a Wiener process with respect to $\mathbf{P}_*$, and

$$d\tilde{S}(t) = \tilde{S}(t)\sigma dw_*(t),$$

$$dS(t) = S(t)[rdt + \sigma dw_*(t)]. \tag{5.14}$$

Theorem 5.36 *The Black–Scholes market is complete.*

Proof. Let $\psi \in \mathcal{L}_2(\Omega, \mathcal{F}_T, \mathbf{P}_*)$ be an arbitrary claim. By the martingale representation theorem (or by Theorem 4.51) applied to the probability space $(\Omega, \mathcal{F}_T, \mathbf{P}_*)$, it follows that there exists a process $f \in \mathcal{L}_{22}$ such that

$$e^{-rT}\psi = e^{-rT}\mathbf{E}_*\psi + \int_0^T f(t)dw_*(t).$$

(We mean the space $\mathcal{L}_{22}$ defined with respect to the measure $\mathbf{P}_*$.) By (5.14), it follows that

$$e^{-rT}\psi = e^{-rT}\mathbf{E}_*\psi + \int_0^T \gamma(t)d\tilde{S}(t), \quad \gamma(t) \triangleq f(t)\tilde{S}(t)^{-1}\sigma^{-1}.$$

Hence the process $\tilde{X}(t) \triangleq e^{-rT}\mathbf{E}_*\psi + \int_0^t \gamma(s)d\tilde{S}(s)$ is the discounted wealth generated by a self-financing strategy such that the quantity of the stock portfolio is $\gamma(t)$. Since $f \in \mathcal{L}_{22}$, it is easy to see that this strategy is admissible. □

Corollary 5.37 *The measure $\mathbf{P}_*$ is the only equivalent risk-neutral measure on $\mathcal{F}_T$.*

Theorem 5.36 does not explain how to calculate the replicating strategy and the corresponding initial wealth. The following theorem gives a method of calculation for an important special case.

Theorem 5.38 *Let functions $\Psi: \mathbf{R} \to \mathbf{R}$ and $\varphi: \mathbf{R} \times [0,T] \to \mathbf{R}$ be such that certain conditions on their smoothness and growth are satisfied (it suffices to require that Ψ and φ are continuous and bounded). Let*

$$\psi = e^{rT}\left[\Psi(\tilde{S}(T)) + \int_0^T \varphi(\tilde{S}(t),t)dt\right].$$

Then this claim is replicable with the initial wealth $\tilde{V}(S(0),0)$ and with the stock quantity $\gamma(t) = \partial(\tilde{V}/\partial x)(\tilde{S}(t),t)$. The corresponding discounted wealth is

$$\tilde{X}(t) = \tilde{V}(\tilde{S}(t),t) + \int_0^t \varphi(\tilde{S}(s),s)ds,$$

where $\tilde{V}$ is the solution of Problem (5.15). In addition,

$$\tilde{X}(t) = \mathbf{E}_*\left\{\Psi(\tilde{S}(T)) + \int_0^T \varphi(\tilde{S}(t),t)dt \,\middle|\, \mathcal{F}_t\right\} \quad \forall t \le T.$$

Moreover,

$$\tilde{V}(\tilde{S}(t),t) = \mathbf{E}_*\left\{\tilde{V}(\tilde{S}(\tau),\tau) + \int_t^\tau \varphi(\tilde{S}(s),s)ds \,\middle|\, \mathcal{F}_t\right\} \quad \forall t,\tau : 0 \le t \le \tau \le T.$$

Proof. Let $\tilde{V}(x,s)$ be the solution of the Cauchy problem for the backward parabolic equation

$$\frac{\partial \tilde{V}}{\partial s}(x,s) + \mathcal{A}\tilde{V}(x,s) = -\varphi(x,s),$$

$$\tilde{V}(x,T) = \Psi(x). \tag{5.15}$$

Here $x > 0$, $s \in [0, T]$,

$$\mathcal{A}v(x) \triangleq \frac{1}{2}\sigma^2 x^2 \frac{d^2 v}{dx^2}(x).$$

Note that the assumptions on Ψ and φ have not yet been specified. Starting from now, we assume that they are such that problem (5.15) has a unique classical solution in the domain $\{(x, s)\} = (0, +\infty) \times [0, T]$.[2]

Let $\tau \in [0, T]$. Similarly to Theorem 4.42, by Ito formula,

$$\begin{aligned}
&\tilde{V}(\tilde{S}(\tau), \tau) - \tilde{V}(S(0), 0) \\
&\quad = \int_0^\tau \left[\frac{\partial \tilde{V}}{\partial t} + \mathcal{A}\tilde{V}\right](\tilde{S}(t), t)dt + \int_0^\tau \frac{\partial \tilde{V}}{\partial x}(\tilde{S}(t), t)d\tilde{S}(t) \\
&\quad = -\int_s^\tau \varphi(\tilde{S}(t), t)dt + \int_s^\tau \frac{\partial \tilde{V}}{\partial x}(y^{x,s}(t), t)d\tilde{S}(t).
\end{aligned}$$

Hence

$$\tilde{V}(\tilde{S}(\tau), \tau) + \int_0^\tau \varphi(\tilde{S}(t), t)dt = \tilde{V}(S(0), 0) + \int_0^T \frac{\partial \tilde{V}}{\partial x}(\tilde{S}(t), t)d\tilde{S}(t).$$

For $\tau = T$, this gives

$$\Psi(\tilde{S}(T)) + \int_0^T \varphi(\tilde{S}(t), t)dt = \tilde{V}(S(0), 0) + \int_0^T \frac{\partial \tilde{V}}{\partial x}(\tilde{S}(t), t)d\tilde{S}(t).$$

Hence this claim is replicable with the initial wealth $\tilde{V}(S(0), 0)$ and with the stock quantity $\gamma(t) = (\partial \tilde{V}/\partial y)(\tilde{S}(t), t)$. The corresponding discounted wealth is

$$\begin{aligned}
\tilde{X}(t) = X(0) + \int_0^t \gamma(t)d\tilde{S}(s) &= X(0) + \int_0^t \frac{\partial \tilde{V}}{\partial y}(\tilde{S}(s), s)d\tilde{S}(s) \\
&= \tilde{V}(\tilde{S}(t), t) + \int_0^t \varphi(\tilde{S}(s), s)ds.
\end{aligned}$$

2 At this point, note that the change of variable x for $y = \ln x$ makes this equation a non-degenerate parabolic equation in the domain $\{(y, s)\} = \mathbf{R} \times [0, T]$. This can help to see which conditions for φ and Ψ are sufficient.

and

$$\tilde{X}(T) = \Psi(\tilde{S}(T)) + \int_0^T \varphi(\tilde{S}(t), t)dt.$$

The corresponding wealth is $X(t) = e^{rt}\tilde{X}(t)$, and the amount of bonds is $\beta(t) = [X(t) - \gamma(t)S(t)]B(t)^{-1}$. This completes the proof. □

Remark 5.39 If the volatility process $\sigma(t)$ is non-random but time dependent and such that $|\sigma(t)| \geq \text{const.} > 0$ and some regularity conditions are satisfied, then the market is also complete. The proof of Theorem 5.38 can be repeated for this case with σ replaced by time-dependent $\sigma(t)$ in the definition for $\mathcal{A}$.

Problem 5.40 Under the assumptions of Theorem 5.38, find an initial wealth and a strategy that replicates the claim $\psi = e^{rT}\Psi(\tilde{S}(T)) = \tilde{S}(T)^{-1}$.

Solution. We need to find $\tilde{V}$ for $\Psi(x) = e^{-rT}x^{-1}$ and $\varphi \equiv 0$. In this case, $\tilde{V}$ can be found explicitly: $\tilde{V}(x,t) = e^{-rT}e^{\sigma^2(T-t)}x^{-1}$ (verify that this $\tilde{V}$ is the solution of (5.15)). Then $(\partial\tilde{V}/\partial x)(x,t) = -e^{-rT}e^{\sigma(T-t)}x^{-2}$, and

$$\gamma(t) = \frac{\partial\tilde{V}}{\partial x}(\tilde{S}(t), t) = -e^{-rT}e^{\sigma^2(T-t)}\tilde{S}(t)^{-2}.$$

The initial wealth is $X(0) = \tilde{V}(S(0), 0) = e^{-rT}e^{\sigma^2 T}S(0)^{-1}$. □

Problem 5.41 Under the assumptions of Theorem 5.38, find an initial wealth and a strategy that replicates the claim $\psi = \tilde{S}(T)^2$.

Solution. We have that $\psi = \tilde{S}(T)^2 = e^{rT}\Psi(\tilde{S}(T))$, where $\Psi(x) = e^{-rT}x^2$. We need to find $\tilde{V}$ for this Ψ and $\varphi \equiv 0$. In this case, $\tilde{V}$ can be found explicitly: $\tilde{V}(x,t) = e^{-rT}e^{\sigma^2(T-t)}x^2$ (verify that this $\tilde{V}$ is the solution of (5.15)). Then $(\partial\tilde{V}/\partial x)(x,t) = 2e^{-rT}e^{\sigma(T-t)}x$, and

$$\gamma(t) = \frac{\partial\tilde{V}}{\partial x}(\tilde{S}(t), t) = 2e^{-rT}e^{\sigma^2(T-t)}\tilde{S}(t).$$

The initial wealth is $X(0) = \tilde{V}(S(0), 0) = e^{-rT}e^{\sigma^2 T}S(0)^2$. □

5.9 Option pricing

5.9.1 Options and their prices

Let us repeat the definitions of the most generic options: the European call option and the European put option. Let terminal time $T > 0$ be given.

A European call option contract traded (contracted and paid) in $t = 0$ is such that the buyer of the contract has the right (not the obligation) to buy one unit of the underlying asset (from the issuer of the option) in $T > 0$ at the strike price K. The market price of option payoff (in T) is $\max(0, S(T) - K)$, where $S(T)$ is the asset price, and K is the strike price.

A European put option contract traded in $t = 0$ gives to the buyer of the contract the right to sell one unit of the underlying asset in $T > 0$ at the strike price K. The market price of option payoff (in T) is $\max(0, K - S(T))$, where $S(T)$ is the asset price, and K is the strike price.

In a more general case, for a given function $F(x) \geq 0$, the European option with payoff $F(S(T))$ can be defined as a contract traded in $t = 0$ such that the buyer of the contract receives an amount of money equal to $F(S(T))$ at time $T > 0$.

In the most general setting, a non-negative function $F : C(0, T) \times C(0, T) \to \mathbf{R}$ is given. Let $\psi = F(S(\cdot), B(\cdot))$. The European option with payoff $F(S(\cdot), B(\cdot))$ is a contract traded in $t = 0$ such that the buyer of the contract receives an amount of money equal to ψ at time $T > 0$.[3]

The following special cases are covered by this setting:

- (vanilla) European call option: $\psi = (S(T) - K)^+$, where $K > 0$ is the strike price;
- (vanilla) European put option: $\psi = (K - S(T))^+$;
- share-or-nothing European call option: $\psi = S(T)\mathbb{I}_{\{S(T)>K\}}$;
- an Asian option: $\psi = f_1\left(\int_0^T f_2(S(t))dt\right)$, where f_i are given functions.

There are many other examples, including exotic options of the European type.

The key role in mathematical finance belongs to a concept of the 'fair price' of options, or derivatives.

The following pricing rule was suggested by Black and Scholes (1973) and has its origins in the model suggested by Bachelier (1900).[4]

Definition 5.42 *The fair price of an option at time $t = 0$ is the minimal initial wealth such that, for any market situation, it can be raised with some acceptable strategies to a wealth such that the option obligation can be fulfilled.*

In fact, we assume that Definition 5.42 is valid for options of all types. We rewrite it now more formally for European options.

3 The options with payoff at given time T are said to be European options. Another important class of options is the class of so-called American options. For these options, the option holder can exercise the option at any time $\tau \in [0, T]$ by his/her choice (see related Definition 3.43; more details are given in Chapter 6).

4 In our notations, the Bachelier model corresponds to the case when $S(t) = S(0) + at + \sigma w(t)$. This model is less popular than the one introduced above, because it gives less realistic distribution of the stock prices: for instance, $S(t)$ may be negative. However, the mathematical properties of this model are very close to the properties of the model introduced above.

Definition 5.43 *The fair price at time $t = 0$ of the European option with payoff ψ is the minimal wealth $X(0)$ such that there exists an admissible self-financing strategy $(\beta(\cdot), \gamma(\cdot))$ such that*

$$X(T) \geq \psi \quad a.s.$$

for the corresponding wealth $X(\cdot)$.

5.9.2 *The fair price is arbitrage-free*

Starting from now and up to the end of this section, we assume that a continuous time market is complete with constant r and σ.

Let us extend the definition of the strategy by assuming that a strategy may include buying and selling bonds, stock and options. Short selling is allowed but all transactions must be self-financing; they represent redistribution of the wealth between different assets. There are no outputs or inputs of wealth. For instance, a trader may borrow an amount of money x to buy k options with payoff ψ at time $t = 0$, then his/her total wealth at time T will be $k\psi - e^{rT}x$. Assume that Definition 5.29 is extended for these strategies.

Proposition 5.44 *Assume that an option seller sells at time $t = 0$ an option with payoff ψ for a price c_+ higher than the fair price c_F of the option. Then he/she can have an arbitrage profit.*

Proof. Assume that the seller has zero initial wealth and sells the option for the price c_+. After that, the seller can invest the wealth $c_+ - c_F > 0$ to bonds, and use the initial wealth $X(0) \triangleq c_F$ with the replicating strategy (that exists) to replicate the claim ψ. Therefore, the option obligation will be fulfilled, and the seller will have a profit equal to $e^{rT}(c_+ - c_F) > 0$. □

Proposition 5.45 *Assume that an option buyer buys at time $t = 0$ an option with payoff ψ for a price c_- that is lower than the fair price c_F of the option. Then he/she can have an arbitrage profit.*

Proof. Assume that the buyer has zero initial wealth. He/she can borrow money and buy the option for the price $c_- < c_F$. The option holder receives the amount of money equal to ψ at time $t = T$ and needs to repay his debt $e^{rT}c_-$ (with the interest), so the resulting wealth is $\psi - e^{rT}c_-$. In addition to this portfolio, the buyer may create the auxiliary portfolio with some self-financing strategy such that the discounted wealth is

$$\tilde{X}(t) = -\int_0^t \gamma(s)d\tilde{S}(s).$$

Here $\gamma(t)$ is the quantity of the stock in the self-financing strategy that replicates the claim ψ with the initial wealth c_F, i.e., $e^{rT}\tilde{X}(T) = -\psi + e^{rT}c_F$. Clearly, $\tilde{X}(0) = 0$. The total wealth for both portfolios is $e^{rT}(c_F - c_-) > 0$. Therefore, the buyer will have a risk-free profit equal to this amount. □

Corollary 5.46 *The fair price of options is the only price that does not allow arbitrage opportunities either for the option seller or for the option buyer.*

Problem 5.47 Is it possible to prove analogs of Propositions 5.44 and 5.45 for the discrete time market model?

5.9.3 Option pricing for a complete market

For a complete market, Definition 5.43 leads to replication.

Theorem 5.48 *Let the market be complete, and let a claim ψ be such that $\mathbf{E}_*\psi^2 < +\infty$, and $\psi = F(S(\cdot))$, where $F(\cdot) : C(0,T) \to \mathbf{R}$ is a function. Then the fair price (from Definition 5.43) of the European option with payoff ψ at time T is*

$$e^{-rT}\mathbf{E}_*\psi, \tag{5.16}$$

and it is the initial wealth $X(0)$ such that there exists an admissible self-financing strategy $(\beta(\cdot), \gamma(\cdot))$ such that

$$X(T) = \psi \quad a.s.$$

for the corresponding wealth.

Proof. From the completeness of the market, it follows that the replicating strategy exists and the corresponding initial wealth is equal to $e^{-rT}\mathbf{E}_*\psi$. Let us show that it is the fair price. Let $X'(0) < c_F$ be another initial wealth, then $\mathbf{E}_*\tilde{X}'(T) = X'(0) < c_F = e^{-rT}\mathbf{E}_*\psi$ for the corresponding discounted wealth $\tilde{X}'(\cdot)$. Hence it cannot be true that $e^{rT}\tilde{X}'(T) \geq \psi$ a.s. □

We shall refer to the price (5.16) as the *Black–Scholes price* of a European option for the case of a complete market.

Corollary 5.49 *Let the assumptions of Theorem 5.38 be satisfied. Let functions $\Psi : \mathbf{R} \to \mathbf{R}$ and $\varphi : \mathbf{R} \times [0,T] \to \mathbf{R}$ be such that certain conditions on their smoothness and growth are satisfied such that problem (5.15) has a unique classical solution in $\{x,s\} = (0,+\infty) \times [0,T]$. Let an option claim ψ be such that*

$$\psi = e^{rT}\left[\Psi(\tilde{S}(T)) + \int_0^T \varphi(\tilde{S}(t), t)dt\right].$$

Then the fair price (Black–Scholes price) of the option at time $t = 0$ is $\tilde{V}(S(0), 0)$, where $\tilde{V}$ is the solution of problem (5.15).

Proof. By Theorem 5.38, $X(0) = \tilde{V}(S(0), 0)$ is the initial wealth for the replicating strategy. □

Corollary 5.50 *The Black–Scholes price does not depend on the appreciation rate $a(\cdot)$.*

Proof. By Theorems 5.48 and 5.38, the fair price is $\tilde{V}(S(0), 0)$, and the equation for $\tilde{V}$ does not include a. Then the proof follows. Another way to prove this corollary is to notice that $\tilde{S}(t) = S(0)\exp(\sigma w_*(t) - \frac{1}{2}\sigma^2 t)$, i.e., the distribution of $\tilde{S}(T)$ under $\mathbf{P}_*$ does not depend on a. □

Corollary 5.51 *(Put–call parity). Let $K > 0$ be given. Let H_p be the Black–Scholes price of the call option with payoff $F_c(S(T)) = (S(T) - K)^+$, and let H_c be the Black–Scholes price of the put option with payoff $F_p(S(T)) = (K - S(T))^+$. Then $H_c - H_p = S(0) - e^{-rT}K$.*

Proof. It suffices to note that $(S(T) - K)^+ - (K - S(T))^+ = S(T) - K$, and $H_c - H_p = e^{-rT}\mathbf{E}_*(S(T) - K) = S(0) - e^{-rT}K$. □

Theorem 5.52 *Consider the Black–Scholes model given volatility σ and the risk-free rate r. Let $\psi = e^{rT}\Psi(\tilde{S}(T))$, where $\Psi : \mathbf{R} \to \mathbf{R}$ is a function. Then the fair price of the option at time $t = 0$ is*

$$e^{-rT}\mathbf{E}_* e^{rT}\Psi(\tilde{S}(T)) = \mathbf{E}_*\Psi(S(0)e^{\eta_0}),$$

where

$$\eta_0 \triangleq -\frac{\sigma^2 T}{2} + \sigma w_*(T).$$

We have that $\eta_0 \sim N(-\sigma^2 T/2, \sigma^2 T)$ under $\mathbf{P}_*$. In other words, the price is

$$\mathbf{E}_*\Psi(\tilde{S}(T)) = \frac{1}{\sqrt{2\pi}}\int_{-\infty}^{+\infty} e^{-\frac{x^2}{2}}\bar{\Psi}\left(S_0 \exp\left[-\frac{\sigma^2 T}{2} + \sigma\sqrt{T}x\right]\right)dx. \quad (5.17)$$

Proof. It suffices to apply the previous result, bearing in mind that $\tilde{S}(T) = S(0)e^{\eta_0}$, $S(T) = e^{rT}\tilde{S}(T)$. □

Corollary 5.53 *The Cox–Ross–Rubinstein market model with an increasing number of periods gives the Black–Scholes price as the limit.*

Proof. The proof for the case when $T = 1$ follows from Corollary 3.57 and Theorem 5.52, since the corresponding formulae for the expectation are identical. For the case when $T \neq 1$, it suffices to note that $\mathrm{Var}\ \ln \tilde{S}(T) = \sigma^2 T$. □

Problem 5.54 Under the assumptions of Theorem 5.38, find the fair price of the option with payoff $\psi = S(T)^2$, i.e., find $e^{-rT}\mathbf{E}_* S(T)^2$.

Solution 1. We have that $\psi = S(T)^2 = e^{rT}\Psi(\tilde{S}(T))$, where $\Psi(x) = e^{rT}x^2$. We need to find the solution $\tilde{V}$ of (5.15) for this Ψ and $\varphi \equiv 0$. In this case, $\tilde{V}$ can be found explicitly: $\tilde{V}(x,t) = e^{rT}e^{\sigma^2(T-t)}x^2$ (verify that this $\tilde{V}$ is the solution of (5.15)). The initial wealth is

$$X(0) = \tilde{V}(S(0), 0) = e^{rT}e^{\sigma^2 T}S(0)^2.$$

In addition, we can obtain the replicating strategy: we have that $(\partial \tilde{V}/\partial x)(x,t) = 2e^{rT}e^{\sigma(T-t)}x$, and

$$\gamma(t) = \frac{\partial \tilde{V}}{\partial x}(\tilde{S}(t), t) = 2e^{rT}e^{\sigma^2(T-t)}\tilde{S}(t).$$

Solution 2. Set $y(t) \triangleq S(t)^2$. We have that

$$\begin{aligned} dy(t) &= 2S(t)dS(t) + S(t)^2\sigma^2 dt = 2S(t)^2[adt + \sigma dw(t)] + S(t)^2\sigma^2 dt \\ &= 2y(t)[adt + \sigma dw(t)] + y(t)\sigma^2 dt = 2y(t)[rdt + \sigma dw_*(t)] + y(t)\sigma^2 dt. \end{aligned}$$

Hence

$$y(t) = y(0) + 2\int_0^t y(s)\sigma dw_*(s) + \int_0^t y(s)[2r + \sigma^2]ds.$$

We have that $y(0) = S(0)^2$.

Set $M(t) \triangleq \mathbf{E}_* y(t) = \mathbf{E}_* S(t)^2$. We have that $M(0) = S(0)^2$,

$$M(t) = M(0) + \int_0^t [2r + \sigma^2]M(s)ds.$$

It follows that $dM(t)/dt = [2r + \sigma^2]M(t)$. Hence $M(t) = e^{[2r+\sigma^2]t}M(0)$. Therefore, $M(T) = \mathbf{E}_* y(T) = \mathbf{E}_* S(T)^2 = e^{[2r+\sigma^2]T}S(0)^2$. Then the price of the option is $e^{-rT}e^{[2r+\sigma^2]T}S(0)^2 = e^{rT}e^{\sigma^2 T}S(0)^2$. □

5.9.4 A code for the fair option price

We give below some examples of codes that illustrate how to apply our formulae in numerical calculation. *This course is not intended to be a course in programming, so all codes provided here are very illustrative and generic, they are not optimal in terms of effectiveness. You may prefer to write your own codes.* For instance, we do not use MATLAB functions for integral or MATLAB *erf* functions.

Example 5.55 Let $(s, T, v, r) = (1, 1, 0.2, 0.07)$. Then the fair price of the option with payoff $F(S(T))$, where $F(x) = 1 + \cos(x)$ is 1.4361. Let $(s, T, v, r) = (2, 1, 0.2, 0.07)$. Then the price of the option is 0.5443.

MATLAB code for the price of an option with payoff $F(x) = 1 + \cos(x)$

```
function[f]=option(s,r,T,v)
N=800; eps=0.01; f=0; pi=3.1415;
for k=1:800;    x=-4+eps*(k-1);
f=f+eps/sqrt(2*pi*T)
*exp(-x^2/(2*T))*(1+cos(s*exp((r-v^2/2)*T+v*x)));
end; f=exp(-r*T)*f;
```

Problem 5.56 (i) Write your own code for calculation of the fair price for payoff $F(S(T))$, where $F(x) = |\sin(4x)|e^x$. (ii) Let $(S(0), T, \sigma, r) = (2, 1, 0.2, 0.07)$. Find the option price with payoff $F(S(T))$.

5.9.5 Black–Scholes formula

We saw already that the fair option price (Black–Scholes price) can be calculated explicitly for some cases. The corresponding explicit formula for the price of European put and call options is called the Black–Scholes formula.

Let $K > 0$, $\sigma > 0$, $r \geq 0$, and $T > 0$ be given. We shall consider two types of options: call and put, with payoff function ψ, where $\psi = (S(T) - K)^+$ or $\psi = (K - S(T))^+$, respectively. Here K is the strike price.

Let $H_{BS,c}(x, K, \sigma, T, r)$ and $H_{BS,p}(x, K, \sigma, T, r)$ denote the fair prices at time $t = 0$ for call and put options with the payoff functions $F(S(T))$ described above given (K, σ, T, r) and under the assumption that $S(0) = x$. Then

$$H_{BS,c}(x, K, \sigma, T, r) = x\Phi(d_+) - Ke^{-rT}\Phi(d_-),$$

$$H_{BS,p}(x, K, \sigma, T, r) = H_{BS,c}(x, K, \sigma, T, r) - x + Ke^{-rT}, \tag{5.18}$$

where

$$\Phi(x) \triangleq \frac{1}{\sqrt{2\pi}} \int_{-\infty}^{x} e^{-\frac{s^2}{2}} ds,$$

and where

$$d_+ \triangleq \frac{\ln(x/K) + Tr}{\sigma\sqrt{T}} + \frac{\sigma\sqrt{T}}{2},$$

$$d_- \triangleq d_+ - \sigma\sqrt{T}. \tag{5.19}$$

This is the celebrated Black–Scholes formula. Note that the formula for put follows from the formula for call from the put–call parity (Corollary 5.51).

Numerical calculation via the Black–Scholes formula

MATLAB code for $\Phi(\cdot)$

```
function[f]=Phi(x)
N=400; eps=abs(x+4)/N; f=0; pi=3.1415;
for k=1:N; y=x-eps*(k-1);
f=f+eps/sqrt(2*pi)*exp(-y^2/2); end;
```

Here $N = 400$ is the number of steps of integration that defines preciseness. One can try different $N = 10, 20, 100, \ldots$. (See also the MATLAB *erf* function.)

MATLAB code for Black–Scholes formula (call)

```
function[x]=call(x,K,v,T,r) x=max(0,s-K);
if T>0.001
d=(log(s/K)+T*(r+v^2/2))/v/sqrt(T);
d1=d-v*sqrt(T);
x=s*Phi(d)-K*exp(-r*T)*Phi(d1); end;
```

MATLAB code for Black–Scholes formula (put)

```
function[x]=put(x,K,v,T,r)
x=call(x,K,v,T,r)-s+K*exp(-r*T); end;
```

Problem 5.57 Assume that $r = 0.05$, $\sigma = 0.07$, $S(0) = 1$. Write a code and calculate the Black–Scholes price of the call option with the strike price $K = 2$ for three months. (*Hint:* three-month term corresponds to $T = 1/4$.)

5.10 Dynamic option price process

In this section, we consider again the Black–Scholes model with non-random volatility σ and non-random risk-free interest rate r.

Definition 5.58 *The fair price of the option at time t is the minimal $\mathcal{F}_t$-measurable random variable (the initial wealth) $X(t)$ such that there exists an admissible* *self-financing strategy* $\{(\beta(s), \gamma(s))\}_{s\in[t,T]}$, *such that*

$$X(T) \geq \psi \quad a.s.$$

for the corresponding wealth.

Theorem 5.59 *The fair price of the option from Definition 5.58 is*

$$\mathbf{E}_*\{e^{-r(T-t)}\psi \mid \mathcal{F}_t\}, \tag{5.20}$$

and it is the wealth $X(t)$ *such that there exists an admissible self-financing strategy* $(\beta(s), \gamma(s))_{s\in[t,T]}$, *such that*

$$X(T) = \psi \quad a.s.$$

for the corresponding wealth.

Proof. Let $(\beta(\cdot), \gamma(\cdot))$ be the replicating strategy that replicates the claim ψ at the time interval $[0, T]$ with the corresponding wealth $X(t)$ and the discounted wealth $\tilde{X}(t)$. We have that

$$e^{-rT}\psi = \tilde{X}(T) = X(0) + \int_0^T \gamma(s)d\tilde{S}(s) = \tilde{X}(t) + \int_t^T \gamma(s)d\tilde{S}(s).$$

Let $\hat{X}(s) = e^{r(T-t)}X(s)$, $\hat{S}(s) = e^{r(T-t)}S(s)$. Then

$$e^{-r(T-t)}\psi = \hat{X}(t) + \int_t^T \gamma(s)d\hat{S}(s). \tag{5.21}$$

Clearly, Theorem 5.16 can be rewritten for the case when $[0, T]$, $\tilde{X}(t)$, and $\tilde{S}(t)$ are replaced for $[t, T]$, $\hat{X}(t)$, and $\hat{S}(t)$. It follows that $(\gamma(s), \beta(s))|_{[t,T]}$ is the replicating strategy on the interval $[t, T]$ with the initial wealth $X(t)$. Further, the theorem statements follow after taking the condition expectation $\mathbf{E}_*\{\cdot \mid \mathcal{F}_t\}$ of both parts of (5.21).

Let us show that $X(t)$ is the fair price. Let $X'(t)$ be another initial wealth at time t such that $\mathbf{P}(X'(t) < X(t)) > 0$, then $\mathbf{E}_*\{\tilde{X}'(T)|\mathcal{F}_t\} = \tilde{X}'(t)$, and $\mathbf{P}(\tilde{X}'(T) < e^{-rT}\mathbf{E}_*\psi \mid \mathcal{F}_t) > 0$, for the corresponding discounted wealth $\tilde{X}'(t) \triangleq e^{-rt}X'(t)$. Hence it cannot be true that $\tilde{X}'(T) \geq \psi$ a.s. □

Definition 5.60 *Let* $c(t)$ *be the price of an option at time* $t \in [0, T]$. *The process* $e^{-rt}c(t)$ *is said to be the discounted price of the option.*

Corollary 5.61 *Let* $\psi = F(S(T))$, *where* $F : \mathbf{R} \to \mathbf{R}$ *is a function. Let the function* $\Psi : \mathbf{R} \to \mathbf{R}$ *be such that* $F(x) = e^{rT}\Psi(e^{-rT}x)$, *i.e.,* $\psi = e^{rT}\Psi(\tilde{S}(T))$.

Then the following holds:

(i) The fair price of the option at time $t \in [0, T)$ depends only on $(S(t), t)$, and it is

$$V(S(t),t) = e^{rt}\hat{V}(\tilde{S}(t),t) = e^{rt}\mathbf{E}_*\{\Psi(\tilde{S}(T))\,|\,\mathcal{F}_t\}. \tag{5.22}$$

Here $\hat{V}$ is the solution of (5.15) for this Ψ and $\varphi \equiv 0$, and

$$V(x,t) = e^{-rt}\hat{V}(e^{rt}x,t). \tag{5.23}$$

(ii) The discounted fair price of the option is

$$\hat{V}(\tilde{S}(t),t) = e^{-rt}V(S(t),t),$$

and it is a martingale with respect to the risk-neutral measure $\mathbf{P}_*$.

Proof. Statement (i) follows from Theorem 5.38 (with $\varphi \equiv 0$). Further, it follows from (5.22) that $\hat{V}(\tilde{S}(t),t) = \mathbf{E}_*\{\Psi(\tilde{S}(T))\,|\,\mathcal{F}_t\}$ for all $t < T$. Then statement (ii) follows. (See, for example, Problem 2.17.) □

Corollary 5.62 *Let $t \in [0, T]$ be given. Under the assumptions of Corollary 5.61, the fair price $V(S(t), t)$ of the option at time t given the current stock price $S(t)$ is such that, for any non-random time $\theta \in (t, T)$,*

$$\tilde{V}(\tilde{S}(t),t) = \mathbf{E}_*\{\tilde{V}(\tilde{S}(\theta),\theta)\,|\,S(t)\},$$

$$V(S(t),t) = e^{-r(\theta-t)}\mathbf{E}_*\{V(S(\theta),\theta)\,|\,S(t)\}, \qquad V(S(T),T) = F(S(T)).$$

In particular,

$$V(S(t),t) = e^{-r\Delta}\mathbf{E}_*\{V(S(t+\Delta),t+\Delta)\,|\,S(t)\}, \qquad \Delta \in [0, T-t].$$

Corollary 5.62 allows us to calculate V consequently starting from $t = T$.

Black–Scholes parabolic equation

By (5.23), we have that $V(t,x) \triangleq e^{-r(T-t)}\mathbf{E}_*\{F(S(T))\,|\,S(t) = x\} = e^{rt}\tilde{V}(e^{-rt}x,t)$, where $\hat{V}$ is the solution of (5.15) with $\varphi \equiv 0$. It follows that $e^{-rt}V(e^{rt}x,t) = \tilde{V}(x,t)$. With this change of the variables, parabolic equation (5.15) is converted to the equation

$$\frac{\partial V}{\partial t}(t,x) + \frac{\sigma^2 x^2}{2}\frac{\partial^2 V}{\partial x^2}(t,x) = r\Big[V(t,x) - x\frac{\partial V}{\partial x}(t,x)\Big],$$

$$V(T,x) = F(x). \tag{5.24}$$

This is the so-called Black–Scholes parabolic equation.

5.11 Non-uniqueness of the equivalent risk-neutral measure

Typically, an equivalent risk-neutral measure is not unique in the case of random volatility (even if it is constant in time). If an equivalent risk-neutral measure is not unique then, by Theorem 5.35, the market cannot be complete, i.e., there are claims ψ that cannot be replicable.

In this section, we assume that r is non-random and constant. Let $\mathcal{F}_t^S$ be the filtration generated by the process $S(t)$. (For this case of non-random r, $\mathcal{F}_t^S \equiv \mathcal{F}_t^{S,r}$.) For the general case, the filtration $\mathcal{F}_t$ generated by the process $(w(t), \eta(t))$ is larger than $\mathcal{F}_t^S$.

5.11.1 *Examples of incomplete markets*

An example with $a \equiv r$

Let $a(t) \equiv r(t)$, and let $\sigma = \sigma(t, \eta)$, where η is a random process (or a random vector, or a random variable), independent from the driving Wiener process $w(t)$ (for instance, η may represent another Wiener process). Clearly, any original probability measure $\mathbf{P} = \mathbf{P}_\eta$ is a risk-neutral measure (note that $\mathcal{Z} \equiv 1$ for any η). Any probability measure is defined by the pair (w, η), therefore it depends on the choice of η. In other words, different η may generate different risk-neutral measures. Clearly, it can happen that two of these different measures are equivalent (it suffices to take two $\eta = \eta_i$, $i = 1, 2$, such that their probability distributions are equivalent, i.e., have the same sets of zero probability). Therefore, an equivalent risk-neutral measure depends on the choice of η, and it may not be unique. It is different from the case of non-random σ.

The simplest example is the following.

Example 5.63 Consider a single stock market model with traded options on that stock. Assume that all option prices are based on the hypothesis that under any risk-neutral measure the volatility is random, independent from time, and can take only two values, σ_1 and σ_2, with probabilities p and $1-p$ respectively, where $p \in [0, 1]$ is given. Let $\mathbf{E}_*$ be the expectation generated by the measure $\mathbf{P}_*$ such that

$$\mathbf{P}_*(A) = p\mathbf{P}_*(A \,|\, \sigma = \sigma_1) + (1-p)\mathbf{P}_*(A \,|\, \sigma = \sigma_2)$$

for all random events A. In this case, the pricing formula $e^{-rT}\mathbf{E}_*\psi$ applied to the price of call option with the strike price K and expiration time T gives

$$\begin{aligned} e^{-rT}\mathbf{E}_* \max(0, S(T) - K) = {} & p\, H_{BS,c}(S(0), K, \sigma_1, r, T) \\ & + (1-p)\, H_{BS,c}(S(0), K, \sigma_2, T, r), \end{aligned}$$

where $H_{BS,c}(S(0), K, \sigma_i, T, r)$ is the Black–Scholes price for call with non-random volatility σ_i given $(S(0), K, r, T)$. Clearly, any $p \in [0, 1]$ defines its own risk-neutral probability measure, and therefore it defines its own $\mathbf{E}_* = \mathbf{E}_*^{(p)}$.

The model in this example can be described as $\sigma(\eta, t) \equiv \eta$, where η is a time-independent random variable that can take only two values, σ_1 and σ_2, with probabilities p and $1-p$ respectively. In this case, any $p \in [0, 1]$ defines the distribution of η, and it defines its own risk-neutral probability measure on $\mathcal{F}_T$ and on $\mathcal{F}_T^S$, and all these measures with $p \in (0, 1)$ are mutually equivalent. Therefore, the equivalent risk-neutral measure is not unique on $\mathcal{F}_T^S$ in this case. In particular, any η defines its own $\mathbf{E}_*\psi$.

Case when $a \neq r$, and (a, σ) are random

Let $(a, \sigma) = f(t, \eta)$, for some deterministic function f and for a random vector η independent from $w(\cdot)$. Formally, any η generates its own risk-neutral measure $\mathbf{P}_*$ defined by Theorem 4.58, since $\mathcal{Z}$ depends on η. It can happen that two of these measures are equivalent but different on $\mathcal{F}_T^S$ (i.e., the equivalent risk-neutral measure is not unique on $\mathcal{F}_T^S$ in this case).

Remark 5.64 Usually, it can be proved that the pricing rule $e^{-rT}\mathbf{E}_*\psi$ is arbitrage-free for a wide selection of $\mathbf{P}_*$. However, the claim $F(S(T))$ is non-replicable for the general case of random volatility.

Case when a is random but σ is non-random

Let $\sigma(t)$ be non-random, and let $a = f(t, \eta)$ for some deterministic function f and for a random vector η that does not depend on time and on $w(\cdot)$. Formally, any η generates again its own risk-neutral measure $\mathbf{P}_*$ defined in Theorem 4.58, since $\mathcal{Z}$ depends on η. However, for any η, we have that $d\tilde{S}(t) = \tilde{S}(t)\sigma(t)dw_*(t)$, where $w_*(\cdot)$ is a process defined in Theorem 4.58. Therefore, the distribution of $\tilde{S}(\cdot)$ and $S(\cdot)$ is uniquely defined by the distribution of $w_*(\cdot)$. Since $w_*(\cdot)$ is a Wiener process under the corresponding measure $\mathbf{P}_*$ for any η, then the distribution of $S(\cdot)$ is the same under all these $\mathbf{P}_*$. In other words, all these measures coincide on $\mathcal{F}_T^S$.

In addition, note that theoretical problems also arise for the case of random r.

5.11.2 *Pricing for an incomplete market*

Mean-variance hedging

Similarly to the case of the discrete time market, Definition 5.43 leads to super-replication for incomplete markets. Clearly, it is not always meaningful. Therefore, there is another popular approach for an incomplete market.

Definition 5.65 *(mean-variance hedging). The fair price of the option is the initial wealth $X(0)$ such that $\mathbf{E}|X(T) - \psi|^2$ is minimal over all admissible self-financing strategies.*

In many cases, this definition leads to the option price $e^{-rT}\mathbf{E}_*\psi$, where $\mathbf{E}_*$ is the expectation for a risk-neutral equivalent measure that needs to be chosen by some optimal way, since this measure is not unique for an incomplete market. This measure needs to be found via solution of an optimization problem. In fact, this method is the latest big step in the development of modern pricing theory. It requires some additional non-trivial analysis outside of our course.

Completion of the market

Sometimes it is possible to make an incomplete market model complete by adding new assets. For instance, if $\sigma(t)$ is random and evolves as the solution of an Ito equation driven by a new Wiener process $W(t)$, then the market can be made complete by allowing trading of any option on this stock (say, European call with given strike price). All other options can be replicated via portfolio strategies that include the stock, the option, and the bond.

A similar approach can be used for the case of random r. Remember that, in our generic setting, we called the risk-free investment a bond, and it was considered as a risk-free investment. In reality, there are many different bonds (or fixed income securities). In fact, they are risky assets, similarly to stocks (discussed in the next section). If r is random, then the market can be made complete by including additional fixed income securities.

5.12 A generalization: multistock markets

Similarly, we can consider a multistock market model, when there are N stocks. Let $\{S_i(t)\}$ be the vector of the stock prices. The most common continuous time model for the prices is again based on Ito equations, which now can be written as

$$dS_i(t) = S_i(t)\left[a_i(t)dt + \sum_{j=1}^{n} \sigma_{ij}(t)dw_j(t)\right], \qquad i = 1, \dots, N.$$

Here $w(t) = (w_1(t), \dots, w_n(t))$ is a vector Wiener process; i.e., its components are scalar Wiener processes. Further, $a(t) = \{a_i(t)\}$ is the vector of the appreciation rates, and $\sigma(t) = \{\sigma_{ij}(t)\}$ is the volatility matrix.

We assume that the components of $w(t)$ are independent.

The equation for the stock prices may be rewritten in the vector form:

$$dS(t) = \mathbf{S}(t)[a(t)dt + \sigma(t)dw(t)],$$

where $S(t) = (S_1(t), \dots, S_N(t))$ is a vector with values in $\mathbf{R}^N$, $\mathbf{S}(t)$ is a diagonal matrix in $\mathbf{R}^{N\times N}$ with the main diagonal $(S_1(t), \dots, S_N(t))$.

Similarly to the case of single stock markets, we assume there is also the risk-free bond or bank account with price $B(t)$ such as described in Section 5.2. In particular, we assume that (5.2) holds, where $r(t)$ is a process of risk-free interest rates that is adapted with respect to the filtration generated by $(w(t), \eta(t))$, where $\eta(t)$ is some random process independent from $w(\cdot)$.

The strategy (portfolio strategy) is a process $(\beta(t), \gamma(t))$ with values in $\mathbf{R} \times \mathbf{R}^N$, $\gamma(t) = (\gamma_1(t), \ldots, \gamma_N(t))$, where $\gamma_i(t)$ is the quantity of the ith stock, and $\beta(t)$ is the quantity of the bond. The total wealth is $X(t) = \beta(t)B(t) + \sum_i \gamma_i(t)S_i(t)$. A strategy $(\beta(\cdot), \gamma(\cdot))$ is said to be *self-financing* if there is no income from or outflow to external sources. In that case,

$$dX(t) = \sum_{i=1}^{N} \gamma_i(t)dS_i(t) + \beta(t)dB(t).$$

To ensure that $S(t)$ and $X(t)$ are well defined as Ito processes, some restrictions on measurability and integrability must be imposed for the processes a, σ, γ, and β.

It can be seen that

$$\beta(t) = \frac{X(t) - \sum_{i=1}^{N} \gamma_i(t)S_i(t)}{B(t)}.$$

Let $\tilde{S}(t) = \{\tilde{S}_i(t)\} \triangleq \exp\left(\int_0^t r(s)ds\right)$ be the discounted stock price. Similarly to Theorem 5.16, it can be shown that

$$d\tilde{X}(t) = \sum_{i=1}^{N} \gamma_i(t)d\tilde{S}_i(t),$$

where $\tilde{X}(t) \triangleq \exp\left(\int_0^t r(s)ds\right)$ is the discounted wealth.

Then absence of arbitrage for this model can be described loosely as the condition that a risk-free gain cannot be achieved with a self-financing strategy.

The following example shows that absence of arbitrage for single stock markets defined for isolated stocks does not guarantee that the corresponding multistock market with the same stocks is arbitrage-free.

Example 5.66 Let $N = 2$, $n = 1$, and let

$$d\tilde{S}_i(t) = \tilde{S}_i(t)[\tilde{a}_i(t)dt + \sigma_i(t)dw(t)],$$

where $\tilde{a}_i(t)$ and $\sigma_i(t)$ are some pathwise continuous processes, $\sigma_2(t) \geq \text{const.} > 0$. Let $\psi(t) \triangleq \sigma_1(t)\tilde{S}_1(t)[\tilde{S}_2(t)\sigma_2(t)]^{-1}$ and $I(t) \triangleq \mathbb{I}_{\{a_1(t) > \psi(t)a_2(t)\}}$,

$$\gamma_1(t) \equiv I(t), \quad \gamma_2(t) \equiv -I(t)\psi(t).$$

Then

$$\begin{aligned} d\tilde{X}(t) = \gamma_1(t)d\tilde{S}_1(t) + \gamma_1(t)d\tilde{S}_2(t) &= I(t)[d\tilde{S}_1(t) - \psi(t)d\tilde{S}_2(t)] \\ &= I(t)[\tilde{a}_1(t) - \psi(t)\tilde{a}_2(t)]dt. \end{aligned}$$

Hence

$$\tilde{X}(T) = X(0) + \int_0^T (\tilde{a}_1(t) - \psi(t)\tilde{a}_2(t))^+ dt.$$

Clearly, this two-stock market model allows arbitrage for some $\tilde{a}_i(\cdot)$. □

In the last example, the model was such that $N > n$. However, it is possible that $N > n$ and the market is still arbitrage-free.

Similarly to the case when $n = m = 1$, it can be shown that the market is arbitrage-free if there exists an equivalent risk-neutral measure such that the discounted stock price vector $\tilde{S}(t)$ is a martingale.

Let us show that there is no arbitrage if there exists a process $\theta(t)$ with values in $\mathbf{R}^m$ such that

$$\sigma(t)\theta(t) = \tilde{a}(t) \tag{5.25}$$

and such that some conditions of integrability of θ are satisfied. (These conditions are always satisfied if the process $\theta(t)$ is bounded.) This process $\theta(t)$ is called the market price of risk process. Here $\tilde{a}(t) \triangleq \tilde{a}(t) - r(t)\mathbf{1}$, where $\mathbf{1} = (1, \ldots, 1)^\top \in \mathbf{R}^N$.

Let us show that the existence of the process $\theta(\cdot)$ implies existence of an equivalent risk-neutral measure. It suffices to show that the measure $\mathbf{P}_*$ defined in the Girsanov theorem (4.58) for this θ is an equivalent risk-neutral measure. Set

$$w_*(t) \triangleq w(t) + \int_0^t \theta(s)ds.$$

By the Girsanov theorem (4.58), $w_*(t)$ is a Wiener process under $\mathbf{P}_*$. Clearly,

$$\sigma(t)dw_*(t) = \tilde{a}(t)dt + \sigma(t)dw(t).$$

In addition, $d\tilde{S}(t) = \tilde{\mathbf{S}}(t)[\tilde{a}(t)dt + \sigma(t)dw(t)]$, where $\tilde{\mathbf{S}}(t)$ is a diagonal matrix with the main diagonal $(\tilde{S}_1(t), \ldots, \tilde{S}_N(t))$. Hence

$$d\tilde{S}(t) = \tilde{\mathbf{S}}(t)\sigma(t)dw_*(t).$$

It follows that $\tilde{S}(t)$ is a martingale under $\mathbf{P}_*$.

For instance, if $n = N$ and the matrix σ is non-degenerate, then $\theta(t) = \sigma(t)^{-1}\tilde{a}(t)$. If this process is bounded, then the market is arbitrage-free.

Problem 5.67 Consider the Black–Scholes market model with stock, bond, and the call options on this stock with the strike prices K_i and expiration times T_i, $i = 1, \ldots, N-1$. These options are priced by the Black–Scholes formula. Consider these options as new risky assets. The new market can be considered as a multistock market model with N stocks ($N - 1$ options plus the original stock). Is this market arbitrage-free? (*Hint:* consider first $N = 2$ and $T_i \geq T$.)

5.13 Bond markets

Bonds are being sold an initial time for a certain price, and the owners are entitled to obtain certain amounts of cash (higher than this initial price) in fixed time (we restrict our consideration to zero-coupon bonds only). Therefore, the owner can have fixed income. Typically, there are many different bonds on the market with different times of maturity, and they are actively traded, so the analysis of bonds is very important for applications.

For the bond-and-stock market models introduced above, we refer to bonds as a risk-free investment similar to a cash account. For instance, it is typical for the Black–Scholes market model where the bank interest rate is supposed to be constant. In reality, the bank interest rate is fluctuating, and its future evolution is unknown. Investments in bonds are such that money is trapped for some time period with a fixed interest rate. Therefore, the investment in bonds may be more or less profitable than the investment in cash account. Thus, there is risk and uncertainty for the bond market that requires stochastic analysis, similarly to the stock market.

General requirements for bond market models

The main features of models for bond markets that generate requirements for the pricing rules are the following:

(i) the process $r(t)$ of bank interest rate is assumed to be random;
(ii) the range for the discounted price processes is bounded;
(iii) the number of securities is larger than the number of driving Wiener processes.

The last feature (iii) has explicit economical sense: there are many different bonds (since bonds with different maturities represent different assets) but their evolution depends on few factors only, and the main factors are the ones that describe the evolution of $r(t)$.

The multistock market model can be used as a model for a market with many different bonds (or fixed income securities). Assume that we are using a multistock market model described above as the model for bonds (i.e., $S_i(t)$ are the bond prices). Feature (iii) can be expressed as the condition that $\sigma_{ij}(t) \equiv 0$ for all $j > n, = 1, \ldots, N$, where n is the number of driving Wiener processes, N is the number of bonds, $N \gg n$. It follows that the matrix σ is degenerate. This is a very essential feature of the bond market. To ensure that the process $\theta(t)$ is finite and the model is arbitrage-free, some special conditions on $\tilde{a}$ must be imposed such that equation (5.25) is solvable with respect to θ. To satisfy these restrictions, the bond market model deals with $\tilde{a}$ being linear functions of σ.

In addition, we have feature (ii): the process $(\tilde{a}, \sigma)$ must be chosen to ensure that the price process is bounded (for instance, $S_i(T) = 1$, $S_i(t) \in [0, 1]$ a.s. if $S_i(t)$ is the price for a zero-coupon bond with the payoff 1 at terminal (maturing) time T).

Consider the case when the bank interest rate $r(t)$ is non-random and known. Let $P(t)$ be the price of a bond with payoff 1 at terminal time T (said to be the maturity time). Clearly, the only price of the bond that does not allow arbitrage for seller and for buyer is

$$P(t) = \exp\left(-\int_t^T r(s)ds\right).$$

In this case, investment in the bond gives the same profit as investment in the cash account. However, this formula cannot be used for the case when $r(s)$ is a random process, since it requires future values of r. In fact, a model for bond prices suggests that the price is

$$P(t) = \mathbf{E}_{Q_t}\left\{\exp\left(-\int_t^T r(s)ds\right)\Big|\,\mathcal{F}_t\right\}, \tag{5.26}$$

where $\mathbf{E}_{Q_t}$ is the expectation generated by a probability measure $\mathbf{Q}_t$, $\mathcal{F}_t$ is the filtration generated by all observable data. The measure $\mathbf{Q}_t$ has to be chosen to satisfy the requirements mentioned above. The choice of this measure may be affected by risk and risk premium associated with particular bonds. (For instance, some bonds are considered more risky than others; to ensure liquidity, they are offered for some lower price, so the possible reward for an investor may be higher.)

Models for bond prices are widely studied in the literature (see the review in Lambertone and Lapeyre, 1996).

An example: a model of the bond market

Let us describe a possible model of a market with N zero-coupon bonds with bond prices $P_k(t)$, where $t \in [0, T_k]$, and where $\{T_k\}_{k=1}^N$ is a given set of maturing times, $T_k \in (0, T]$, $P(T_k, T_k) = 1$.

We consider the case where there is a driving n-dimensional Wiener process $w(t)$. Let $\mathcal{F}_t$ be a filtration generated by this Wiener process. We assume that the process $r(t)$ is adapted to $\mathcal{F}_t$. (To cover some special models, we do not assume that $r(t) \geq 0$.) In addition, we assume that we are given an $\mathcal{F}_t$-adapted and bounded process $q(t)$ that takes values in $\mathbf{R}^n$.

Set the bond prices as

$$P_k(t)=\mathbf{E}\left\{\exp\left(-\int_t^{T_k} r(s)ds+\int_t^{T_k} q(s)^\top dw(s)-\frac{1}{2}\int_t^{T_k} |q(s)|^2 ds\right)\Big|\,\mathcal{F}_t\right\}. \tag{5.27}$$

In this model, different bonds are defined by their maturity times.

Clearly, the processes $P_k(t)$ are adapted to $\mathcal{F}_t$, $P_k(T_k) = 1$, $P_k(0) \in [0, 1]$, and $\tilde{P}_k(t) \triangleq P_k(t)\exp(-\int_0^t r(s)ds) \in [0, 1]$ a.s. In addition, it can be seen that (5.26) holds for the measure $\mathbf{Q}_t = \mathbf{Q}_{t,k}$ such that $\mathbf{Q}_{t,k}/d\mathbf{P} = Z_k(t)$, where

$$Z_k(t) \triangleq \exp\left(\int_t^{T_k} q(s)^\top dw(s) - \frac{1}{2}\int_t^{T_k} |q(s)|^2 ds\right).$$

Theorem 5.68 *Pricing rule (5.27) ensures that, for any k, there exists an $\mathcal{F}_t$-adapted process $\sigma_k(t)$ with values in $\mathbf{R}^n$ such that*

$$d_t P_k(t) = P_k(t)\Big(\Big[r(t) - \sigma_k(t)^\top q(t)\Big]dt + \sigma_k(t)^\top dw(t)\Big), \quad t < T_k. \qquad (5.28)$$

Proof. Let k be fixed. We have that

$$P_k(t) = y(t)z(t)\exp\Big(\int_0^t r(s)ds\Big), \qquad (5.29)$$

where

$$y(t) \triangleq \mathbf{E}\Big\{\exp\Big(-\int_0^{T_k} r(s)ds + \int_0^{T_k} q(s)^\top dw(s) - \frac{1}{2}\int_0^{T_k} |q(s)|^2 ds\Big)\Big| \mathcal{F}_t\Big\},$$

$$z(t) \triangleq \exp\Big(-\int_0^t q(s)^\top dw(s) + \frac{1}{2}\int_0^t |q(s)|^2 ds\Big).$$

It follows from the Clark theorem (4.51) that there exists a square integrable n-dimensional $\mathcal{F}_t$-adapted process $\hat{y}(t) = \hat{y}(t, T_k)$ with values in $\mathbf{R}^n$ such that

$$y(T) = \mathbf{E}y(T) + \int_0^T \hat{y}(t)^\top dw(t).$$

Note that $y(t) > 0$. Set $\delta_k(t) = \hat{y}(t)/y(t)$. Then

$$y(T) = \mathbf{E}y(T) + \int_0^T y(t)\delta_k(t)^\top dw(t).$$

By the Ito formula, it follows that

$$dz(t) = z(t)\Big(|q(t)|^2 dt - q(t)^\top dw(t)\Big).$$

Set $\sigma_k(t) \triangleq \delta_k(t) - q(t)$. Finally, the Ito formula applied to (5.29) implies that (5.28) holds. This completes the proof. □

It follows from (5.28) that this bond market is a special case of the multi-stock market described above, when $S_k(t) = P_k(t)$, $k = 1, \dots, N$, where $\tilde{a}(t) \equiv (\tilde{a}_1(t), \dots, \tilde{a}_N(t))^\top \in \mathbf{R}^N$, and where $\sigma(t)$ is a matrix process with values in $\mathbf{R}^{N\times n}$ such that its kth row is zero for $t > T_k$ and it is equal to $\sigma_k(t)^\top$ for $t \le T_k$. The process $\tilde{a}(t)$ is such that

$$\tilde{a}_k(t) = \begin{cases} -\sigma_k(t)^\top q(t), & t \le T_k \\ 0, & t > T_k. \end{cases}$$

Then the corresponding market price of risk process $\theta(t)$ is $\theta(t) \equiv -q(t)$. This process $\theta(t)$ is bounded if $q(t)$ is bounded, since $\theta(t) \equiv -q(t)$. Note that the case $N \gg n$ is allowed, and the bond market is still arbitrage-free.

To derive an explicit equation for $P_k(t)$ and $\sigma_k(t)$, we need to specify a model for the evolution of the process $(r(t), q(t))$. The choice of this model defines the model for the bond prices. For instance, let $n = 1$, let the process q be constant, and let $r(t)$ be an Ornstein–Uhlenbek process described in Problem 4.41. Then this case corresponds to the so-called Vasicek model (see, e.g., Lambertone and Lapeyre (1996, p. 127)). In this case, $P_k(t)$ can be found explicitly from (5.27).

5.14 Conclusions

- Continuous time models allow explicit and complete solution for many theoretical problems, and they are the main models in mathematical finance.
- A continuous time market model is complete for a generic case of non-random constant volatility (this case can be considered as a limit for the Cox–Ross–Rubinstein model with increasing frequency).
- Continuous time models allow the solution of many pricing problems via Kolmogorov partial differential equations.
- A continuous time market model is based on Ito calculus, and it needs some interpretation to be implemented for a real market with time series of prices. Strategies developed for this model cannot be applied immediately, because they include Ito processes that are not explicitly presented in real market data. For instance, there is the question of how to extract the appreciation rate $a(t)$ and volatility $\sigma(t)$ from the time series of prices.

Interpretation of historical data in view of continuous time models is studied in Chapter 9.

5.15 Problems

Below, a is the appreciation rate, σ is the volatility, r is the risk-free rate, $S(t)$ is the stock price, $X(t)$ is the wealth, $\tilde{S}(t)$ is the discounted stock price, $\tilde{X}(t)$ is the discounted wealth, $(\beta(\cdot), \gamma(\cdot))$ is a self-financing strategy, where γ is the quantity of the stock shares, β is the quantity of the bonds.

Self-financing strategies for a continuous time market

Problem 5.69 Let $r(t) \equiv 0$. Let $\gamma(t) = 1$, $t \in [0, t_1)$, $\gamma(t) = -1/2$, $t \in [t_1, T]$, where $0 \le t_1 < T$. Let $S(0) = 1$, $S(t_1) = 1.1$, $S(T) = 0.95$. Find $\tilde{X}(T)$ and $X(T)$.

Problem 5.70 Solve the previous problem for the case when $r(t) \equiv 0.05$.

Problem 5.71 Let $r(t) \equiv r$, $a(t) \equiv a$, $\sigma(t) \equiv \sigma > 0$ be constant. Let $\gamma(t) = (a - r)\tilde{S}(t)^{-1}$. Find $\mathbf{E}\tilde{X}(T)$.

Problem 5.72 Let $r(t) \equiv r$, $a(t) \equiv a$, $\sigma(t) \equiv \sigma > 0$ be constant. Let $\gamma(t) = k\tilde{X}(t)\tilde{S}(t)^{-1}$, where $\tilde{X}(t)$ is the corresponding wealth, $k \in \mathbf{R}$. Find $\tilde{X}(T)$. (*Hint:* derive a closed Ito equation for $\tilde{X}(t)$ and use the known solutions of these equations.)

Problem 5.73 Let there exist t_1, t_2: $0 \le t_1 < t_2 \le T$ such that $\sigma(t) = 0$, $\tilde{a}(t) \neq 0$, $t \in (t_1, t_2)$ a.s. Prove that this market model allows arbitrage. (*Hint:* take $\gamma(t) = \tilde{a}(t)\mathbb{I}_{t\in(t_1,t_2)}$.)

Claim replication and option price

Assume that (a, r, σ, r) is non-random and constant, and they are given. Assume that $S(0) > 0$ and $T > 0$ are also given.

Problem 5.74 Under the assumptions of Theorem 5.38, find an initial wealth and a strategy that replicates the claim $\psi = e^{rT}\Psi(\tilde{S}(T)) = \tilde{S}(T)^{-2}$. Find the option price for this claim.

Problem 5.75 Find an initial wealth and a strategy that replicates the claim $\psi = (1/T)\int_0^T \tilde{S}(T)dt$. (*Hint:* use Theorem 5.38, and find the solution V of the boundary value Problem 5.15 from the proof of Theorem 5.38 for $\varphi(x, t) = e^{-rT}x/T$, $\Psi \equiv 0$; V can be found explicitly.)

Problem 5.76 Find an initial wealth and a strategy that replicates the claim $\psi = (1/T)\int_0^T S(T)dt$. (*Hint:* use Theorem 5.38 and find the solution V of the boundary value problem (5.15) from the proof of Theorem 5.38 for $\varphi(x, t) = e^{-rT}e^{rt}x/T$, $\Psi \equiv 0$; V can be found explicitly as $V(x, t) = \int_0^T \varphi(x, t)dt$.)

Problem 5.77 Consider an option with payoff $\psi = \mathbb{I}_{\{S(T)>K\}}$, where $K > 0$ (it is the so-called digital option). Express the option price via an integral with the probability density function of a Gaussian random variable.

Problem 5.78 (Bachelier's model). Consider a market model where the risk-free rate $r \ge 0$ is constant and known, and where the stock price evolves as

$$dP(t) = adt + \sigma dw(t),$$

where $\sigma > 0$ is a given constant, $a \in \mathbf{R}$, $w(t)$ is a Wiener process. Assume that the fair price for a call option is $e^{-rT}\mathbf{E}_* \max(P(T) - K, 0)$, where K is the strike price, T is termination time. Here $\mathbf{E}_*$ is the expectation defined by the risk-neutral probability measure (this measure gives the same probability distribution of $P(\cdot)$ as the original measure for the case when $r = a$). Find an analogue of the Black–Scholes formula for the call option. (*Hint:* (1) find the expectation via calculation of an integral with a certain (known) probability density; (2) for simplicity, you may take first $r = 0$.)

Black–Scholes formula

Problem 5.79 Let $H_{BS,c}(s, K, r, T, \sigma)$ and $H_{BS,p}(s, K, r, T, \sigma)$ be the Black–Scholes prices for call and put options respectively. Here $\sigma \in \mathbf{R}$ is the volatility, $r \geq 0$ is the bank interest rate, $s = S(0)$ is the initial stock price, K is the strike price.

(i) Are these functions increasing (decreasing) in s? Prove. (*Hint:* use the basic risk-neutral valuation rule.)
(ii) Find the limits for these functions as: (a) $T \to +\infty$; (b) $\sigma \to +\infty$; (c) $T \to +0$; (d) $\sigma \to +0$. (*Hint:* use the Black–Scholes formula.)

Challenging problems

Problem 5.80 Let $\tilde{a}(t) \equiv \hat{a} \neq 0$ not depend on time, and let a self-financing strategy be defined in closed-loop form such that $\gamma(t) \triangleq k\hat{a}(K - \tilde{X}(t))\tilde{S}(t)^{-1}$, where $\tilde{X}(t)$ is the corresponding discounted wealth. Here $k > 0$ and $K > 0$ are given constants. Prove that $\mathbf{E}|\tilde{X}(t) - K|^2 \to 0$ as $t \to +\infty$. In addition, prove that if $X(0) > K$, then $\tilde{X}(t) > K$ for all $t > 0$ a.s., and if $X(0) < K$, then $\tilde{X}(t) < K$ for all $t > 0$ a.s. Is $\tilde{X}(t)$ bounded from below?

Problem 5.81 Let $X(0) = 1$, and let a self-financing strategy be defined in closed-loop form such that $\gamma(t) \triangleq k(\tilde{X}(t) - 2)\tilde{X}(t)\tilde{S}(t)^{-1}$, where $\tilde{X}(t)$ is the corresponding discounted wealth. Here $k \in \mathbf{R}$ is a given constant. Prove that $\tilde{X}(t) \in (0, 2)$ for all $t > 0$ a.s. In addition, prove that if $\sigma(t) \equiv \sigma > 0$ is constant, then, for any $\varepsilon > 0$, $\int_s^{s+1} \mathbf{P}_*(\varepsilon \leq \tilde{X}(t) \leq 2 - \varepsilon)dt \to 0$ as $s \to +\infty$.

Problem 5.82 Let $\tilde{a}(t) \equiv \hat{a} \neq 0$ not depend on time, and let a self-financing strategy be defined in closed-loop form such that $\gamma(t) \triangleq k\hat{a}(K(t) - \tilde{X}(t))\tilde{S}(t)^{-1}$, where $\tilde{X}(t)$ is the corresponding discounted wealth. Here $k > 0$ is a given constant, $K(t) > 0$ is a given deterministic function bounded in $t > 0$ together with its derivative $dK(t)/dt$. Let $X(0) > K(0)$. Is it possible that $\tilde{X}(t) > K$ for all $t > 0$ a.s.? Is it possible that if $X(0) < K$, then $\tilde{X}(t) < K$ for all $t > 0$ a.s.? Investigate the properties of the process $\tilde{X}(t) - K(t)$ for $t \to +\infty$.

Problem 5.83 (see Example 5.12). Let $\sigma > 0$ and a be constant, let $T = +\infty$, and let $\tau = \min\{t > 0 \colon S(t) = K\}$, where $K \neq S(0)$. Prove that $\mathbf{P}(\tau < +\infty) = 1$ and $\mathbf{E}\tau = +\infty$. (*Hint:* the last equality is easier to prove for $a = r = 0$.)

For simplicity, you can assume that $S(t) = S(0) + w(t)$; it does not remove the main challenge for the previous problem as well as for the following problem.

Problem 5.84 John's initial wealth is $X(0) = S(0)$, and he uses the following strategy: $\gamma(t) = \mathbb{I}_{\{S(t) \geq S(0)\}}$, where $\mathbb{I}$ denotes the indicator function. (This means

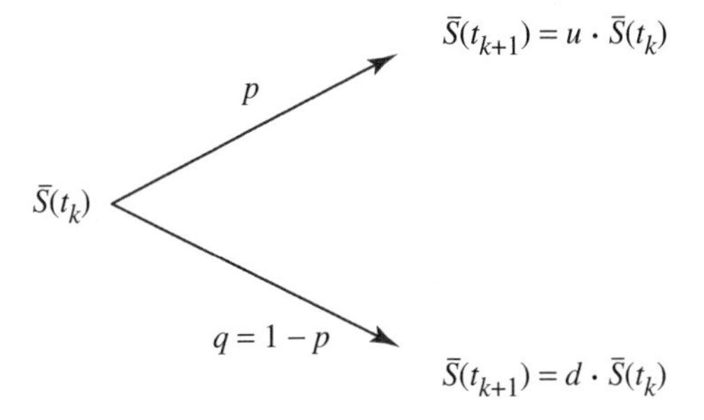

Figure 6.1 Price evolution in the binomial model.

When binomial trees are used in practice, the life of the option is typically divided into a large enough number of steps to ensure good approximation. With 20 time steps, $2^{20} > 10^6$ stock price paths are possible. However, if one chose $u = 1/d$, then $ud = 1$, and the number of possible price paths will be less. Figure 6.2 gives an example of a tree for five steps if $ud = 1$.

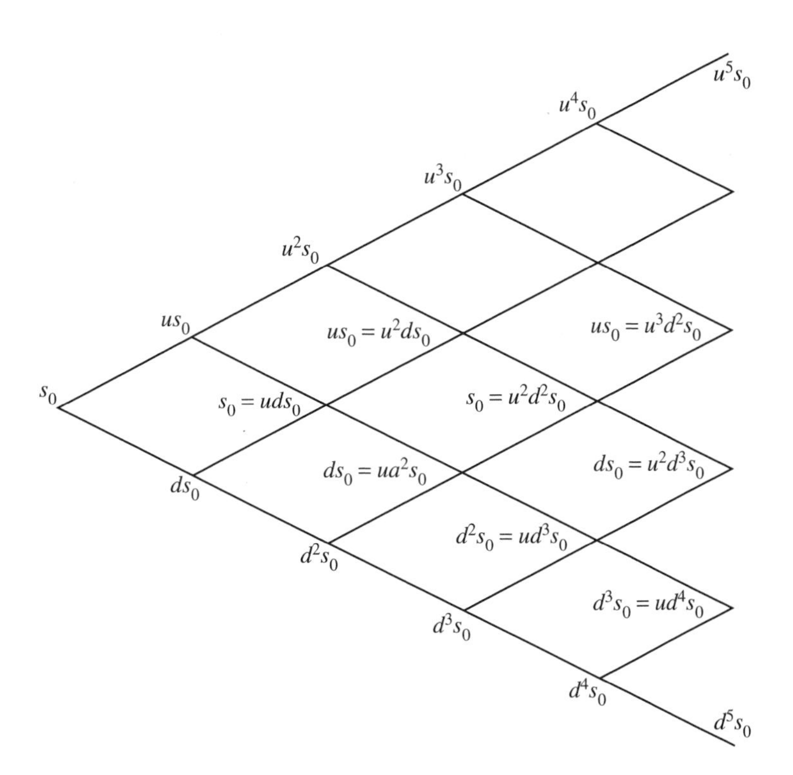

Figure 6.2 Binomial tree for $N = 5$ when $ud = 1$.

Note that higher multinomial trees, for example a trinomial tree, are also widely used.

6.1.2 Choice of u, d, p for the case of constant r and σ

The most popular binomial tree represents the Cox–Ross–Rubinstein discrete time market model described in Section 3.9. It is known as the Cox–Ross–Rubinstein binomial tree.

Let us assume that the continuous time stock price process $S(t)$ is the solution of the Ito equation

$$dS(t) = S(t)(adt + \sigma dw(t)), \tag{6.1}$$

where σ and a are constants. We assumed also that the risk-free interest rate r is constant and non-random.

It was shown above that the Cox, Ross, and Rubinstein model ensures approximation of the continuous time Black–Scholes model for a large number of periods. This means that the options prices calculated via binomial trees approximate the fair prices.

Let $\mathbf{P}_*$ be the risk-neutral equivalent measure. For the Black–Scholes market model with constant non-random (σ, a, r), this measure exists and it is uniquely defined (see Theorem 4.58). (Remember that this measure coincides with the original probability measure iff $a = r$.) Let $\mathbf{E}_*$ be the corresponding conditional expectation.

Let $\mathcal{F}_t$ be the filtration generated by $w(s)$, $s \le t$. (Note that $\mathcal{F}_t$ is also the filtration generated by $S(t)$, and it is also the filtration generated by $\tilde{S}(t)$, or by $w(t)$.) Under the measure $\mathbf{P}_*$, the discounted price process $\tilde{S}(t) \triangleq e^{-rt}S(t)$ is a martingale with respect to $\mathcal{F}_t$.

We shall model the probabilistic characteristics of the price evolution under the risk-neutral measure $\mathbf{P}_*$, as required by the Black–Scholes approach. This means that

$$S(t_{k+1}) = S(t_k)M_{k+1},$$

where

$$\begin{aligned} M_{k+1} &\triangleq e^{r(t_{k+1}-t_k)} \exp\left\{-[t_{k+1} - t_k]\frac{\sigma^2}{2} + \sigma[w_*(t_{k+1}) - w_*(t_k)]\right\} \\ &= e^{a\Delta} \exp\left\{-\Delta\frac{\sigma^2}{2} + \sigma\sqrt{\Delta}\xi_{k+1}\right\}. \end{aligned}$$

Here $w_*(t)$ is a Wiener process under $\mathbf{P}_*$, and ξ_k are i.i.d. (independent identically distributed) random variables with law $N(0, 1)$ under $\mathbf{P}_*$.

Proposition 6.1 *Let $\gamma \sim N(0,\sigma)$. Then $\mathbf{E}e^{\gamma} = e^{\sigma^2/2}$ and $\mathbf{E}e^{-\sigma^2/2+\gamma} = 1$.*

Problem 6.2 Prove Proposition 6.1.

Let us describe our choice of parameters for the binomial tree. Let s_k be the prices at time $t = t_k$ modelled by the binomial tree.

It is natural to assume that

$$s_0 = S(0), \qquad s_{k+1} = s_k \hat{M}_{k+1}, \quad k = 0, 1, 2, \dots,$$

where $\hat{M}_k$ are random variables such that

$$\hat{M}_{t_k} \triangleq \begin{cases} u & \text{with probability } p \\ d & \text{with probability } q = 1 - p. \end{cases}$$

We want the discrete time approximation to be close in some sense to the original continuous stock price. There are three unknown parameters (u, d, p), hence three restrictions can be satisfied. The most popular choice of the restrictions is

$$ud = 1, \quad \mathbf{E}_*\hat{M}_k = \mathbf{E}M_k, \quad \text{Var } \ln \hat{M}_k = \text{Var } \ln M_k. \tag{6.2}$$

We have that

$$\mathbf{E}_* M_{k+1} = e^{r\Delta}, \quad \text{Var } \ln M_k = \sigma^2 \Delta.$$

(The first equation here follows from Proposition 6.1.) Hence we have the following restriction for parameters:

$$\mathbf{E}_*\hat{M}_k = up + d(1-p) = e^{r\Delta},$$

or

$$p = \frac{e^{r\Delta} - d}{u - d}.$$

Further, we need to choose u to ensure that $\text{Var } \ln \hat{M}_k = \text{Var } \ln M_k$ with $d = 1/u$. It leads to the following rule.

Rule 6.3 *(choice of parameters for the binomial tree). The most popular choice of parameters is*

$$u \triangleq e^{\sigma\sqrt{\Delta}}, \quad d \triangleq \frac{1}{u} = e^{-\sigma\sqrt{\Delta}},$$

$$p = \frac{e^{r\Delta} - d}{u - d}, \quad q = 1 - p.$$

In that case, (6.2) holds.

For example, this rule is used in MATLAB Financial Toolbox (Version 2).

Problem 6.4 Prove that the equalities for the variances in (6.2) hold.

Note that the market with the prices (ρ_k, s_k), where $\rho_k = B(0)e^{r\Delta k}$, is a special case of the Cox–Ross–Rubinstein model introduced in Section 3.9, and $d = \rho(1+d_1)$, $u = \rho(1+d_2)$, where d_i are parameters from Section 3.9.

6.1.3 Pricing of European options via a binomial tree

Let us apply the binomial tree described above to pricing of European options. It was shown above that the Black–Scholes formula gives an explicit formula for European put and call (under certain assumptions). Moreover, formula (5.17) gives a good enough solution for a European option general payoff function $F(S(T))$. In addition, it will now be demonstrated how to apply the binomial trees for European options, and then this method will be extended for American options.

Rule 6.5 *For a European call option with payoff $F(S(T))$ the option prices at times t_k can be estimated as $V(s_k, t_k)$, and these values can be calculated backward starting from $t_N = T$, and*

$$V(s_N, t_N) = F(s_N),$$

$$V(s_{k-1}, t_{k-1}) = e^{-r\Delta}\left[pV(u\,s_{k-1}, t_k) + (1-p)V(d\,s_{k-1}, t_k)\right], \tag{6.3}$$

$$k = N, N-1, \dots, 1.$$

Here $s_k = \bar{S}(t_k)$, $t_k = k\Delta$, $\Delta = T/N$. The price at time $t_0 = 0$ is $V(s_0, 0)$, where $s_0 = S(0)$.

(Compare with Corollary 5.62.)

Note that the parameters d, u, p do not depend on the type of option (i.e., on $F(\cdot)$).

Figure 6.3 shows an example of a tree for the stock prices and the corresponding tree for the call option prices.

6.2 American option and non-arbitrage prices

We describe the American option in the continuous time setting. However, all definitions given below are valid in the discrete time setting as well (see Definition 3.43).

Up to the end of this chapter, we assume that the risk-free rate $r \geq 0$ is a non-random constant, and that the process $(a(t), \sigma(t))$ is adapted to the filtration $\mathcal{F}_t$, and $\mathcal{F}_t$, $a(t)$, $\sigma(t)$ are as described in Section 5.1.

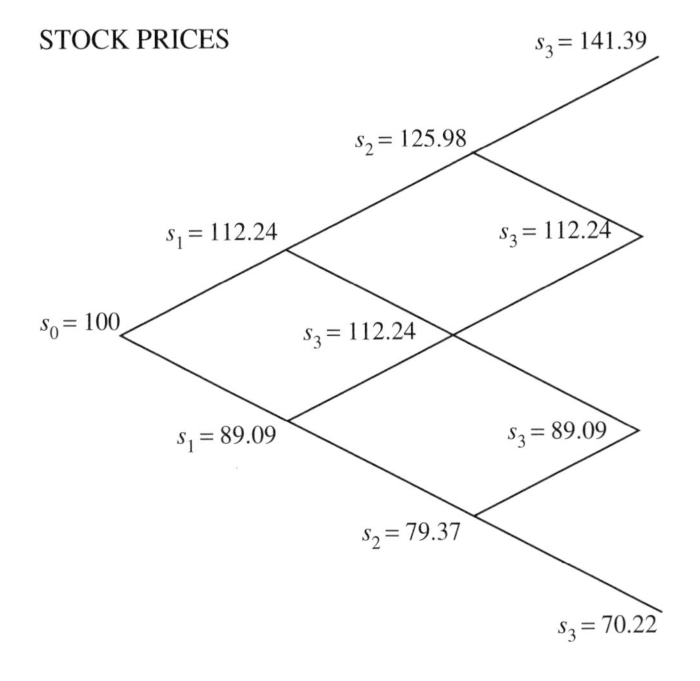

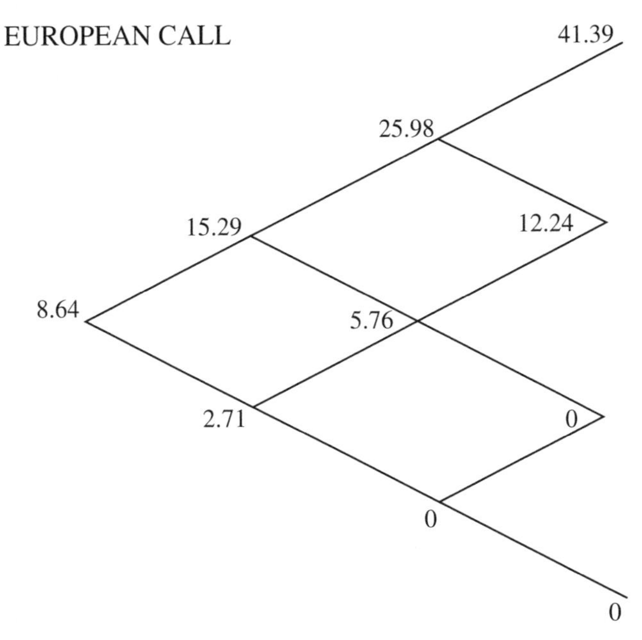

Figure 6.3 Binomial tree for the stock prices and for the call option prices. It is assumed that $S(0) = 100$, $T = 1$, $\sigma = 0.2$, $N = 3$, $r = 0$, $K = 100$ (the strike price). Note that the tree for the prices does not depend on r, and $u = 1.12$, $d = 0.89$, $p = 0.47$.

We assume that all Markov times mentioned below are Markov times with respect to this $\mathcal{F}_t$.

Definitions

An American call (put) option contract traded at time t may stipulate that the buyer (holder) of the contract has the right (not the obligation) to buy (sell) one unit of the underlying asset (from the writer, or issuer of the option) at any time $s \in [t, T]$ (by his/her choice) at the strike price K. The option payoff (in s) is $\max(0, S(s) - K)$ for call and $\max(0, K - S(s))$ for put, where $S(s)$ is the asset price.

More generally, an American option with payoff $F(S(s))|_{s\in[t,T]}$ can be defined as a contract traded at time t that stipulates that the buyer of the contract receives an amount of money equal to $F(S(s))$ at any time $s \in [t, T]$ by his/her choice.

We assume that $F(x)$ is a given function such that $F(x) \geq 0$.

If the option holder has to fulfil the option obligations at time τ, we say that he or she exercises the option, and τ is called the exercise time.

Sub(super)martingale properties for non-arbitrage prices

Similarly to Section 5.9, we shall use the extended definition of the strategy assuming that a strategy may include buying and selling bonds, stock, and options. Short selling is allowed for stocks and bonds and not allowed for options. All transactions must be self-financing; they represent redistribution of the wealth between different assets. For instance, a trader may borrow an amount of money x to buy k American options at time t with payoff $F(S(s))|_{s\in[t,T]}$, then his/her total wealth at exercise time s is $kF(S(s)) - e^{r(s-t)}x$. We assume that Definition 5.29 is extended for these strategies. In addition, we assume that the wealth and strategies are defined for Markov random initial times θ such that $\theta \in [0, T]$. All definitions can be easily rewritten for this case.

In particular, let $X(0)$ be an initial wealth, and let $\tilde{X}(t)$ be the discounted wealth generated by an admissible self-financing strategy $(\beta(\cdot), \gamma(\cdot))$. For any Markov time τ such that $\tau \in [0, T]$, we have

$$\mathbf{E}_*\tilde{X}(\tau) = X(0) + \mathbf{E}_* \int_0^\tau \gamma(t) d\tilde{S}(t)$$

$$= X(0) + \mathbf{E}_* \int_0^\tau \gamma(t)\tilde{S}(t)^{-1}[\tilde{a}(t)dt + \sigma(t)dw(t)] = X(0),$$

since $a(t)dt + \sigma(t)dw(t) = \sigma(t)dw_*(t)$, where $w_*(t) \triangleq w(t) + \int_0^t \sigma(s)^{-1}\tilde{a}(s)ds$ is a Wiener process with respect to the risk-neutral measure $\mathbf{P}_*$, $\tilde{a}(t) \triangleq a(t) - r$.

Similarly, let $X(\theta)$ be an initial wealth at Markov initial time θ such that $\theta \in [0, T]$, and let $\tilde{X}(t)$ be the discounted wealth generated by an admissible self-financing strategy $(\beta(\cdot), \gamma(\cdot))$. For any Markov time τ such that $\tau \in [\theta, T]$, we have

$$\begin{aligned}\mathbf{E}_*\{\tilde{X}(\tau)|\mathcal{F}_\theta\} &= \tilde{X}(\theta) + \mathbf{E}_*\left\{\int_\theta^\tau \gamma(t)d\tilde{S}(t)\Big|\mathcal{F}_\theta\right\} \\ &= X(\theta) + \mathbf{E}_*\left\{\int_0^\tau \gamma(t)\tilde{S}(t)^{-1}dw_*(t)\Big|\mathcal{F}_\theta\right\} = X(\theta).\end{aligned}$$

Lemma 6.6 *Let $C(t)$ be the market price at time t for an American option with the payoff $F(S(s))|_{s\in[t,T]}$ (this price is not necessarily the fair price). We assume that this process is adapted to $\mathcal{F}_t$. Assume that the market does not allow arbitrage for the option buyer. Then*

$$\tilde{C}(\theta) \geq \mathbf{E}_*\{\tilde{C}(\tau)|\mathcal{F}_\theta\}$$

for any two Markov times θ and τ such that $\theta \leq \tau \leq T$ a.s. In other words, $\tilde{C}(t)$ is a supermartingale with respect to the risk-neutral measure $\mathbf{P}_*$.

Proof. Let the lemma's statement be untrue. Then there exists a Markov time θ such that $\theta \leq \tau$ and

$$\tilde{C}(\theta) + \psi = \mathbf{E}_*\{\tilde{C}(\tau)|\mathcal{F}_\theta\},$$

where ψ is an $\mathcal{F}_\theta$-measurable random variable such that $\psi \geq 0$ a.s., and $\mathbf{P}_*(\psi > 0) > 0$. Assume that the buyer has zero initial wealth. He/she buys the option at time θ for the market price $C(\theta) = e^{r\theta}\tilde{C}(\theta)$. In addition, he/she invests $e^{r\theta}\psi$ in the bonds. To obtain cash for these purchases, he/she creates an auxiliary portfolio with negative initial wealth $X(\theta) = -\mathbf{E}_*\{\tilde{C}(\tau)|\mathcal{F}_\theta\}$ and with some self-financing strategy such that the discounted wealth is such that $\tilde{X}(\tau) = -\tilde{C}(\tau)$ (this wealth and strategy exist since the market is complete). Note that the initial wealth for the combined portfolio is zero. At time τ, the buyer sells the option for the market price $C(\tau)$, and this is sufficient to cover the debt generated by the second portfolio that has the total wealth $-C(\tau)$ at this time. The remaining wealth originated from the investment of the amount $e^{r\theta}\psi$ in bonds gives the arbitrage profit. □

Lemma 6.7 *Let $C(t)$ and $\tilde{C}(t)$ be the processes described in Lemma 6.6. Let $\tilde{C}(t) \triangleq e^{-rt}C(t)$ be the discounted option price. Let τ be a Markov time such that the option is not exercised at $s \in [t,\tau)$. Assume that the market does not allow arbitrage for the option seller. Then*

$$\tilde{C}(\theta) \leq \mathbf{E}_*\{\tilde{C}(\tau)|\mathcal{F}_\theta\}$$

for any Markov time θ such that $\theta \in [t,\tau)$. In other words, $\tilde{C}(s)$ is a submartingale on time interval $[t,\tau)$ with respect to the risk-neutral measure $\mathbf{P}_*$.

Proof. Let the lemma's statement be untrue. This means that there exists a Markov time θ such that $\theta \in [t, \tau)$ and

$$\tilde{C}(\theta) = \mathbf{E}_*\{\tilde{C}(\tau)|\mathcal{F}_\theta\} + \eta,$$

where η is an $\mathcal{F}_\theta$-measurable random variable such that $\eta \geq 0$ a.s., and $\mathbf{P}_*(\eta > 0) > 0$. Assume that the seller sells the option at time θ for the market price $C(\theta) = e^{r\theta}\tilde{C}(\theta)$. At this time, he/she invests the wealth $e^{r\theta}\eta$ to bonds, and starts the portfolio with the self-financing strategy such that the corresponding discounted wealth is such that $\tilde{X}(\theta) = \mathbf{E}_*\{\tilde{C}(\tau)|\mathcal{F}_\theta\}$ and $\tilde{X}(\tau) = \tilde{C}(\tau)$. Remember that the option is not exercised before τ. At time τ, the seller is using the wealth $e^{r\tau}\tilde{X}(\tau) = C(\tau)$ to buy back the option from the buyer (or, equivalently, to buy the same option and keep it up to the time when the buyer exercises the option). This allows to fulfil obligations for the option that had been sold. The remaining wealth originated from the investment of $e^{r\theta}\eta$ in bonds gives the arbitrage profit. □

Corollary 6.8 *Let $C(t)$ and $\tilde{C}(t)$ be the processes described in Lemma 6.6. Assume that the market does not allow arbitrage for the option buyer and for the option seller. In that case, $\tilde{C}(t)$ is a supermartingale on $[0, T]$ under the measure $\mathbf{P}_*$. Further, let θ and τ be Markov times such that $\theta \leq \tau$ and the option is not exercised at $[\theta, \tau)$. Then*

$$\tilde{C}(\theta) = \mathbf{E}_*\{\tilde{C}(\tau)|\mathcal{F}_\theta\}$$

for any Markov time θ such that $\theta \leq \tau$ a.s. In other words, $\tilde{C}(t)$ is a martingale on time interval $[\theta, \tau)$ with respect to the risk-neutral measure $\mathbf{P}_*$.

6.3 Fair price of the American option

The following pricing rule is similar to the rule for European options from Definitions 5.42 and 5.43.

Definition 6.9 *The fair price of an American option at time t is the minimal initial wealth $X(t)$ such that, for any market situation, this wealth can be raised with some acceptable strategies to the wealth that can cover option obligations.*

We now rewrite Definition 6.9 more formally.

Definition 6.10 *The fair price at time t of an American option with payoff $F(S(s))|_{s\in[t,T]}$ is the minimal $\mathcal{F}_t$-measurable random variable $c_F(t)$ such that, for any stock price scenario, there exists an admissible self-financing strategy $\{(\beta(s), \gamma(s))\}_{s\in[t,T]}$ such that*

$$X(s) \geq F(S(s)) \quad a.s.$$

for the corresponding wealth $X(s)$ and for all times s such that $s \in [t, T]$.

The fair price does not allow arbitrage

Lemma 6.11 *Assume that an option seller sells at time t an American option with the payoff* $F(S(s))|_{s\in[t,T]}$ *for a price* $c_+(t)$ *such that* $c_+(t) > c_F(t)$, *where* $c_F(t)$ *is the fair price of this option. Then he/she can have an arbitrage profit.*

Proof. Assume that the seller has zero initial wealth and sells the option for the price $c_+(t)$. After that, the seller can invest the wealth $c_+(t) - c_F(t) > 0$ to bonds, and create an additional portfolio with the initial wealth $X(t) \stackrel{\Delta}{=} c_F(t)$ and with an admissible self-financing strategy (that exists) such that $X(s) \geq F(S(s))$ a.s. for all $s \in [t, T]$. Therefore, the option obligation will be fulfilled, and the seller will have the risk-free (arbitrage) profit equal to $e^{r(T-t)}(c_+(t) - c_F(t)) > 0$. □

Lemma 6.12 *Assume that an option seller sells at time t an American option with the payoff* $F(S(s))|_{s\in[t,T]}$ *for a price* $c_-(t)$ *such that* $c_-(t) \leq c_F(t)$, *where* $c_F(t)$ *is the fair price of this option. Then he/she cannot have an arbitrage profit.*

Proof. Assume that the initial wealth of the option seller is zero. At time t, he/she sells the American option for the price $c_-(t)$, and creates a portfolio using this amount of cash as the initial wealth. Let $X(s)$ be the wealth obtained by an admissible self-financing strategy $(\beta(s), \gamma(s))|_{s\in[t,T]}$ with this initial wealth $X(t) = c_-(t)$. By the definition of the fair price, it follows that there exists time $\theta \in [t, T]$ such that $\mathbf{P}(X(\theta) < F(S(\theta))) > 0$ (this time may depend on the strategy). If the option holder exercises the option at time θ, then the seller has losses. Therefore, arbitrage gain for the seller is impossible for any admissible self-financing strategy. Then the proof follows. □

Corollary 6.13 *The fair price of options is the only price that does not allow arbitrage opportunities for the seller and such that the option obligations may be fulfilled with probability 1 for the wealth generated from this price with a self-financing strategy.*

Lemma 6.14 *Assume that the market price of an American option with the payoff* $F(S(s))|_{s\in[t,T]}$ *is the fair price* $c_F(t)$. *Then a buyer of the option cannot have an arbitrage profit.*

Proof. Assume that the option buyer has the initial wealth $c_F(t)$ at time t. He/she buys the option, so his/her total wealth at time $s \geq t$ is either $F(S(s))$ (if the option is exercised) or $c_F(s)$ (if it is being sold).

Assume first that the option is exercised. By the definition of the fair price, there exists an admissible self-financing strategy $\Gamma_t = (\beta(\cdot), \gamma(\cdot))|_{[t,T]}$ such that $X(s) \geq F(s)$ for the corresponding total wealth $X(s)$ such that the initial wealth is $X(t) = c_F(t)$. The strategy Γ_t does not allow arbitrage, and the wealth $X(s)$ is the same or bigger than the buyer's wealth (obtained after the exercise of the option). Therefore, the buyer cannot have arbitrage profit if the option is exercised.

Further, we have that $X(s) \geq c_F(s)$, since the initial wealth $X(t) = c_F(t)$ must ensure that $X(q) \geq F(S(q))$ for all $q \geq t$, and the initial wealth $X(s,s) = c_F(s)$ must ensure that $X(q,s) \geq F(S(q))$ for $q \geq s$ only, where $X(q,s)$ is the total wealth for the self-financing strategy that can cover the option's obligations after time s. Hence $X(s) \geq X(s,s) = c_F(s)$. Again, the strategy Γ_t does not allow arbitrage, and its wealth at time s is the same or bigger than the buyer's wealth. Therefore, the buyer cannot have arbitrage profit at time s. □

Corollary 6.15 *Let $c_F(t)$ be the fair price of the American option, and let $\tilde{c}_F(t) \triangleq e^{-rt}c_F(t)$ be the discounted fair price. Then*

(i) $\tilde{c}_F(t)$ is a supermartingale for $t \in [0,T]$;
(ii) Let θ and τ be Markov times such that $\theta \leq \tau$ and the option is not exercised at $[\theta,\tau)$. Then $\tilde{c}_F(\rho) = \mathbf{E}_*\{\tilde{c}_F(\tau)|\mathcal{F}_\rho\}$ for any Markov time ρ such that $\rho \in [\theta,\tau]$. In other words, $\tilde{c}_F(t)$ is a martingale on time interval $[\theta,\tau)$ with respect to the risk-neutral measure $\mathbf{P}_*$.

Proof follows from Corollary 6.8 and Lemmas 6.6 and 6.7. □

6.4 The basic rule for the American option

Up to the end of this chapter, we assume that the market is complete with constant $r \geq 0$ and $\sigma > 0$.

In addition, we assume that $\mathcal{F}_t$ is the filtration generated by $S(t)$ (i.e., by $w(t)$, or by $\tilde{S}(t)$).

Let the function $F(x) \geq 0$ be such that some technical conditions are satisfied (it suffices to require that $F(x)$ is continuous and $|F(x)| \leq \text{const.}\ (|x|+1)$). For instance, the functions $F(x) = (x-K)^+$ and $F(x) = (K-x)^+$ are admissible.

Theorem 6.16 *Let initial time $t \in [0,T]$, the risk-free rate $r \geq 0$ and volatility $\sigma > 0$ be given. Consider an American option with the payoff $F(S(s))|_{s\in[t,T]}$. Let $c_F(t)$ be the fair price of this option at time $t \in [0,T]$. Then the following holds:*

(i) $c_F(T) \equiv F(S(T))$;
(ii) $c_F(t) \geq F(S(t))$ for all $t \in [0,T]$, and there exists (random) Markov time $\tau = \tau(\omega) = \tau(t,\omega)$ such that $\tau(\omega) \in [t,T]$ for all ω and

$$c_F(s) > F(S(s)) \quad \forall t \in [t,\tau), \qquad c_F(\tau) = F(S(\tau)).$$

(If $c_F(t) = F(S(t))$, then $\tau = t$.) Moreover, the option will not be exercised at time $[t,\tau)$ by a rational investor.

(iii) $c_F(0) = \mathbf{E}_*e^{-r\tau}F(S(\tau))$.
(iv) The fair price $c_F(0)$ is the solution of the optimal stopping problem

$$c_F(0) = \sup_{\hat{\tau}} \mathbf{E}_*e^{-r\hat{\tau}}F(S(\hat{\tau})), \tag{6.4}$$

where the supremum is over all (random) Markov times $\hat{\tau} = \hat{\tau}(\omega)$ such that $\hat{\tau}(\omega) \in [0, T]$ for all ω. Time τ from (ii) is the optimal stopping time for this problem.

The process $c_F(t)$ and random time τ are uniquely defined up to equivalency, i.e., $\mathbf{P}(\tau = \tau') = 1$ for any other optimal τ'.

In fact, Theorem 6.16 (iv) defines the fair price for an American option uniquely for a wide class of $F(\cdot)$ but it is not an explicit formula. An explicit solution is still unknown. Theorem 6.16 (iv) states that the optimal exercise time τ for an option's buyer is the solution of the optimal stopping problem, when τ is the optimal stopping time. It can be shown that this optimal time τ is the first time that $S(t)$ hits a certain optimal level, $\Gamma(t)$. Therefore, the problem can be reduced to calculation of this $\Gamma(t)$. An explicit solution is unknown even when r and σ are known constants. A possible and most popular numerical solution is described below; it is based on binomial trees. There is much literature devoted to approximate methods of solution; the pricing of American options is not easy.

Proof of Theorem 6.16. Statement (i) is obvious. Let us prove statement (ii). Assume that $c_F(t) < F(S(t))$. Then an option buyer can exercise the option immediately at time t of its purchase and obtain the risk-free gain. It contradicts Lemma 6.14. Hence $c_F(t) \geq F(S(t))$.

Further, if the holder exercises the option when $c_F(s) > F(S(s))$, then he/she obtains the value $F(S(t))$ which is less than the market price $c_F(s)$ of the option. Clearly, he or she would prefer to sell the option rather than exercise it. It follows that τ is the first reasonable time that the option can be exercised. This τ can be found as the first time before T when $c_F(\tau) = F(S(\tau))$; if this does not occur before T, then $\tau = T$, because of (i). It follows that this τ is a Markov time, since it is constructed as a time of first achievement for the currently observable process. Then (ii) follows.

Further, let $t = 0$ be initial time. Let $X(0)$ be the initial wealth equal to the option price $c_F(0)$. We have that

$$\tilde{X}(\tau) = X(0) + \int_0^{\hat{\tau}} \gamma(t) d\tilde{S}(t)$$

for any Markov time $\hat{\tau} \in [0, T]$ and for any self-financing admissible strategy $(\beta(\cdot), \gamma(\cdot))$ with the corresponding discounted wealth $\tilde{X}(t)$. The option writer needs to obtain the total wealth $X(t)$ such that $X(\tilde{\tau}) \geq F(S(\tilde{\tau}))$ for all possible exercise times $\tilde{\tau} \in [0, T]$, i.e.,

$$\tilde{X}(\hat{\tau}) \geq e^{-r\hat{\tau}} F(S(\hat{\tau})), \quad \mathbf{E}_* \tilde{X}(\hat{\tau}) \geq \mathbf{E}_* e^{-r\hat{\tau}} F(S(\hat{\tau})). \tag{6.5}$$

For Markov times $\hat{\tau}$, it follows that

$$X(0) = \mathbf{E}_* \tilde{X}(\hat{\tau}) \geq \mathbf{E}_* e^{-r\hat{\tau}} F(S(\hat{\tau})).$$

By (ii), it follows that τ is a Markov time. By Lemma 6.12, it follows that the option seller cannot have an arbitrage profit if the price is fair. By Lemma 6.7, it follows that

$$X(0) = c_F(0) \leq \mathbf{E}_* e^{-r\tau} c_F(\tau) = \mathbf{E}_* e^{-r\tau} F(S(\tau)).$$

Remember that (6.5) holds for all Markov times $\hat{\tau}$. Hence

$$X(0) = c_F(0) = \mathbf{E}_* e^{-r\tau} F(S(\tau)),$$

and

$$\mathbf{E}_* e^{-r\tau} F(S(\tau)) \geq \mathbf{E}_* e^{-r\hat{\tau}} F(S(\hat{\tau}))$$

for all Markov times $\hat{\tau} \in [0, T]$. Then (iii) and (iv) follow. □

Remark 6.17 By equation (ii) from Theorem 6.16, it follows that the price for an American option cannot be less than the price of the corresponding European option which is $\mathbf{E}_* e^{-r\tau} F(S(\tau))$ with fixed $\tau \equiv T$. This can also be seen from the economical meaning of conditions for both options: the American option gives more opportunities to its holder.

Markov property and price at time $t > 0$

Clearly, all definitions can be rewritten for a market model such that time interval $[0, T]$ is replaced for time interval $[t, T]$, where $t \in (0, T)$ is non-random, and where the initial stock price $S(t)$ is non-random. The complete analogue of Theorem 6.16 (iii)–(iv) is valid (the proof is similar). By this theorem (rewritten for the market with new time interval), the price at time s is

$$c_F(t) = \sup_{\tau} \mathbf{E}_* e^{-r\tau} F(S(\tau)),$$

where supremum is taken over Markov times τ such that $\tau(\omega) \in [t, T]$ for all ω.

Formally, we cannot apply this result directly to the original model for the case when $t > 0$, because $S(t)$ is random and $c_F(t)$ is an $\mathcal{F}_t$-measurable random variable. However, we can use that $S(t)$ is a Markov process under $\mathbf{P}_*$ (see Remark 4.35). It follows that

$$\mathbf{E}_*\{c_F(s) \,|\, \mathcal{F}_t\} = \mathbf{E}_*\{c_F(s) \,|\, S(t)\}, \quad s \geq t.$$

In particular, $c_F(t) = \mathbf{E}_*\{c_F(t) \,|\, S(t)\}$, since $c_F(t)$ is $\mathcal{F}_t$-measurable.

Now observe that the following is satisfied:

(a) The conditional probability space with the probability $\mathbf{P}_*(\cdot \,|\, S(t))$ (i.e., under the condition that $S(t)$ is known) is such that Theorem 6.16 (iii)–(iv) (rewritten for the market with time interval $[t, T]$) can be applied (by the reasons described above).

(b) Let $c_{[t,T]}(s)$ denote the option fair price for the corresponding market at time $s \in [t, T]$, on the conditional probability space with the probability $\mathbf{P}_*(\cdot \mid S(t))$ (i.e., under the condition that $S(t)$ is known). Then $c_{[s,T]}(s) = c_{[0,T]}(s)$ (remember that the first value here is defined on the conditional probability space given $S(s)$).

These observations lead to the following useful addition to Theorem 6.16.

Theorem 6.18 *Under the assumptions and notations of Theorem 6.16, the following holds:*

(i) $c_F(t) = \mathbf{E}_*\{e^{-r(\tau-t)}F(S(\tau))|S(t)\}$, *where* $\tau = \tau(t,\omega)$ *is as described in Theorem 6.16 (ii);*

(ii) The fair price $c_F(t)$ *is the solution of the optimal stopping problem*

$$c_F(t) = \sup_{\hat{\tau}} \mathbf{E}_*\{e^{-r(\hat{\tau}-t)}F(S(\hat{\tau}))\,|S(t)\},$$

where the supremum is over all (random) Markov times $\hat{\tau} = \hat{\tau}(\omega)$ *such that* $\hat{\tau}(\omega) \in [t, T]$ *for all* ω. *Time* τ *is the optimal stopping time for this problem.*

Corollary 6.19 *Under the assumptions and notations of Theorem 6.18, it follows that time* τ *is the best time for the option holder to exercise the option in the following sense:*

$$\mathbf{E}_*X_\tau(T) \geq \mathbf{E}_*X_{\hat{\tau}}(T),$$

for all (random) Markov times $\hat{\tau} = \hat{\tau}(\omega)$ *such that* $\hat{\tau}(\omega) \in [t, T]$ *for all* ω. *Here* $X_{\hat{\tau}}(T)$ *denotes the terminal wealth for the option buyer if he/she exercises the option at time* $\hat{\tau}$, *obtains the amount of cash* $F(S(\hat{\tau}))$, *and invests this amount in the bonds (or in the bank account with interest rate* r*).*

Proof of Corollary 6.19. By Theorem 6.18 (ii), it follows that time τ is such that

$$\mathbf{E}_*e^{-r\tau}F(S(\tau)) \geq \mathbf{E}_*e^{-r\hat{\tau}}F(S(\hat{\tau})),$$

for all (random) Markov times $\hat{\tau} = \hat{\tau}(\omega)$ such that $\hat{\tau}(\omega) \in [0, T]$ for all ω. Clearly, $X_{\hat{\tau}}(T) = e^{r(T-\hat{\tau})}F(S(\cdot))$. Hence

$$\begin{aligned}\mathbf{E}_*X_\tau(T) &= \mathbf{E}_*e^{r(T-\tau)}F(S(\tau)) = e^{rT}\mathbf{E}_*e^{-r\tau}F(S(\tau)) \\ &\geq e^{rT}\mathbf{E}_*e^{-r\hat{\tau}}F(S(\hat{\tau})) = \mathbf{E}_*X_{\hat{\tau}}(T).\end{aligned}$$

This completes the proof. □

The fair price at random time

Theorem 6.16 (iii) can be generalized as the following.

Theorem 6.20 *Let $c_F(t)$ be the fair option at time $t \in [0, T]$. Then, under the assumptions and notations of Theorem 6.16,*

$$e^{-r\theta} c_F(\theta) = \mathbf{E}_*\{e^{-r\tau} F(S(\tau)) \,|\, \mathcal{F}_\theta\}$$

for any Markov time θ such that $\theta \leq \tau$.

Proof. Note that the statement of the theorem for $\theta = 0$ follows from Theorem 6.16. Let us extend the corresponding proof for the general case of random θ.

Let the initial wealth $X(\theta)$ at Markov time θ be equal to the option price $c_F(\theta)$. We have that

$$\tilde{X}(\tau) = \tilde{X}(\theta) + \int_\theta^{\hat{\tau}} \gamma(t) d\tilde{S}(t)$$

for any Markov time $\hat{\tau} \in [\theta, T]$ and for any self-financing admissible strategy $(\beta(\cdot), \gamma(\cdot))$ with the corresponding discounted wealth $\tilde{X}(t)$. The option writer needs to obtain the total wealth $X(\cdot)$ such that $X(\tilde{\tau}) \geq F(S(\tilde{\tau}))$ for all possible exercise times $\tilde{\tau} \in [\theta, T]$, i.e.,

$$\tilde{X}(\hat{\tau}) \geq e^{-r\hat{\tau}} F(S(\hat{\tau})), \quad \tilde{X}(\theta) = \mathbf{E}_*\{\tilde{X}(\hat{\tau})|\mathcal{F}_\theta\} \geq \mathbf{E}_*\{e^{-r\hat{\tau}} F(S(\hat{\tau}))|\mathcal{F}_\theta\}. \tag{6.6}$$

Here $\tilde{X}(t) = e^{-rt}X(t)$. For Markov times $\hat{\tau}$, it follows that

$$\tilde{X}(\theta) = \mathbf{E}_*\{\tilde{X}(\hat{\tau})|\mathcal{F}_\theta\} \geq \mathbf{E}_*\{e^{-r\hat{\tau}} F(S(\hat{\tau}))|\mathcal{F}_\theta\}.$$

Remember that τ is a Markov time. By Lemma 6.12, it follows that the option seller cannot have an arbitrage profit for the fair price. By Lemma 6.7, it follows that

$$\tilde{X}(\theta) \leq \mathbf{E}_*\{e^{-r\tau} c_F(\tau)|\mathcal{F}_\theta\} = \mathbf{E}_*\{e^{-r\tau} F(S(\tau))|\mathcal{F}_\theta\}.$$

Remember that (6.6) holds for all Markov times $\hat{\tau}$. Hence

$$\tilde{X}(\theta) = e^{-r\theta} c_F(\theta) = \mathbf{E}_*\{e^{-r\tau} F(S(\tau))|\mathcal{F}_\theta\}.$$

Then the proof follows. □

The following theorem establishes some causality property for the price of the American option: the price at a given time can be represented via the price at a later time.

Theorem 6.21 *Under the assumptions and notations of Theorem 6.20,*

$$e^{-r\theta} c_F(\theta) = \mathbf{E}_* \left\{ e^{-r\psi} c_F(\psi) \,\middle|\, \mathcal{F}_\theta \right\}$$

for all Markov times ψ and θ such that $\theta \le \psi \le \tau$ a.s.

Proof follows from Theorem 6.20. □

6.5 When American and European options have the same price

Note that it is possible that the Black–Scholes price of a European put option at time t with the strike price K is less than $(K - S(t))^+$ (to see this, solve Problem 5.79 (i) for put with $r > 0$). In contrast, it is impossible for the American put, so the price for the American put can be higher than the price of the corresponding European put. The question arises as to whether there are cases when American and European options have the same price.

Theorem 6.22 *(Merton's theorem). Consider an American call option, i.e., with the payoff $F(S(t)) \equiv (K - S(t))^+$. Then the fair price for this option is the same as for the corresponding European call.*

Proof. We only consider the case of $t = 0$, the other cases being similar. Since $\tilde{S}(t) \triangleq S(t)e^{-rt}$ is a martingale under $\mathbf{P}_*$, for any Markov time taking values in $[0, T]$, we have that $\tilde{S}(\tau) = \mathbf{E}_*\{\tilde{S}(T)|\mathcal{F}_\tau\}$, and

$$\tilde{S}(\tau) - e^{-rT}K = \mathbf{E}_*\{e^{-rT}(S(T) - K)|\mathcal{F}_\tau\} \le \mathbf{E}_*\{e^{-rT}(S(T) - K)^+|\mathcal{F}_\tau\}.$$

Clearly,

$$e^{-r\tau}(S(\tau) - K)^+ = (\tilde{S}(\tau) - e^{-r\tau}K)^+ \le (\tilde{S}(\tau) - e^{-rT}K)^+.$$

Hence

$$e^{-r\tau}(S(\tau) - K)^+ \le \mathbf{E}_*\{e^{-rT}(S(T) - K)^+|\mathcal{F}_\tau\}.$$

By taking expectations, we obtain

$$\mathbf{E}_* e^{-r\tau}(S(\tau) - K)^+ \le \mathbf{E}_* e^{-rT}(S(T) - K)^+ = e^{-rT}\mathbf{E}_*(S(T) - K)^+,$$

and it is the Black–Scholes price. By Remark 6.17, it follows that the fair price of the American option cannot be less then the Black–Scholes price. Then the proof follows. □

The previous theorem can be generalized as the following.

Theorem 6.23 *Consider an American option with the payoff* $F(S(s))|_{s\in[t,T]}$ *such that the function* $F(x)$ *is convex in* $x > 0$*, and such that the function* $\alpha F(x/\alpha)$ *is non-increasing in* $\alpha \in (0, 1]$*. Then, under the notations of Theorem 6.16,* $\tau \equiv T$*, and the fair price is the same for American and European calls.*

Proof. We only consider the case of $t = 0$; the other cases are similar. It follows from the assumptions about $F(\cdot)$ that, for any Markov time taking values in $[0, T]$,

$$e^{-r\tau}F(S(\tau)) = e^{-r\tau}F(e^{r\tau}\tilde{S}(\tau)) \leq e^{-rT}F(e^{rT}\tilde{S}(\tau)).$$

Since $\tilde{S}(t) \triangleq S(t)e^{-rt}$ is a martingale under $\mathbf{P}_*$, we have $\tilde{S}(\tau) = \mathbf{E}_*\{\tilde{S}(T)|\mathcal{F}_\tau\}$. In addition, let us assume that $\mathbf{P}(\tau < T) > 0$. We have that the support of the conditional distribution of $S(z)$ given $S(\theta)$ is $(0, +\infty)$. Since $F(\cdot)$ is convex and non-linear, it follows from Jensen's inequality[1] that

$$e^{-rT}F(e^{rT}\tilde{S}(\tau)) < e^{-rT}\mathbf{E}_*\{F(e^{rT}\tilde{S}(T))|\mathcal{F}_\tau\} = \mathbf{E}_*\{e^{-rT}F(S(T))|\mathcal{F}_\tau\}.$$

Hence

$$e^{-r\tau}F(S(\tau)) < \mathbf{E}_*\{e^{-rT}F(S(T))|\mathcal{F}_\tau\}.$$

By taking expectations, we obtain

$$\mathbf{E}_*e^{-r\tau}F(S(\tau)) < \mathbf{E}_*e^{-rT}F(S(T)) = e^{-rT}\mathbf{E}_*F(S(T)),$$

i.e., it is the fair price of the European option. Therefore, the supremum over all τ is achieved only for $\tau = T$. Then the proof follows. □

Note that the function $F(x) = (x - K)^+$ (for the American call option) is such that the assumptions of Theorem 6.23 are satisfied.

Theorem 6.24 *Consider an American option with convex function* $F(x)$*, for a market with* $r = 0$ *(i.e., with zero risk-free interest rate). Then, under the notations of Theorem 6.16,* $\tau \equiv T$*, and the fair price is the same for American and European options with this* $F(\cdot)$*.*

Problem 6.25 (challenging problem). Prove Theorem 6.24. (*Hint:* use Jensen's inequality.)

1 Let $[a, b] \subset \mathbf{R}$ be given, $-\infty \leq a < b \leq +\infty$. Let $f : [a, b] \to \mathbf{R}$ be a function that is convex in $x \in [a, b]$. Let ξ be an integrable random variable such that $f(\xi)$ is also integrable. Then $f(\mathbf{E}\xi) \leq \mathbf{E}f(\xi)$. (This is *Jensen's inequality*.)

Corollary 6.26 *Note that the American put option has convex payoff function $F(x) = \max(K - x, 0)$. Then, under the notations of Theorem 6.16, $\tau \equiv T$, and the fair price is the same for the American and European put options, if $r = 0$. (In addition, it was proved already that the fair price is the same for American and European call options for $r \geq 0$.)*

However, the prices for the corresponding American and European put options are different if $r > 0$. Thus, we need a numerical calculation algorithm for the fair prices of American put options for $r > 0$.

6.6 Stefan problem for the price of American options

In this section, we assume that the function $F(x)$ is absolutely continuous and such that $|F(x)| + |dF(x)/dx| \leq \text{const}$.

By Theorem 6.18 (ii), it follows that the solution $c_F(t)$ of the optimal stopping problem (6.6) can be represented as $c_F(t) = V(S(t), t)$, where

$$V(x,t) = \sup_{\tau} \mathbf{E}_*\{e^{-r\tau} F(S(\tau)) \,|\, S(t) = x\}, \tag{6.7}$$

and where supremum is taken over Markov times τ such that $\tau(\omega) \in [t, T]$ for all ω.

By Theorem 6.21 (i)–(ii), it follows that $V(x, T) \equiv F(x)$ and $V(x, s) \geq F(x)$.

Let

$$D \triangleq \{(x,t) : \ V(x,t) > F(x), \ \ t \in [0,T]\}.$$

Let $\tilde{V}(x,t) = e^{-rt} V(e^{rt}x, t)$, i.e., $V(x,t) = e^{rt}\tilde{V}(e^{-rt}x, t)$. Then $\tilde{c}_F(t) = \tilde{V}(\tilde{S}(t), t)$.

Let us assume that the solution V of optimal stopping problem (6.7) is continuous and locally bounded in D together with the derivatives V'_t, V'_x and V''_{xx}.[2] By the Ito formula, it follows that

$$d\tilde{c}_F(t) = \left[\frac{\partial \tilde{V}}{\partial t}(\tilde{S}(t), t) + \frac{\sigma^2 \tilde{S}(t)^2}{2}\frac{\partial^2 \tilde{V}}{\partial x^2}(\tilde{S}(t), t)\right]dt + \frac{\partial \tilde{V}}{\partial x}(\tilde{S}(t), t)\tilde{S}(t)\sigma\, dw_*(t)$$

when t evolves inside a connected time interval such that $c_F(t) \in D$. (Remember that $w_*(t) = w(t) + \int_0^t \sigma^{-1} a(s) ds$ and $\tilde{S}(t)\sigma\, dw_*(t) = d\tilde{S}(t)$.) On the other hand, $\tilde{c}_F(t)$ is a martingale under $\mathbf{P}_*$ for this time interval (Corollary 6.15 (ii)). It follows that

$$\frac{\partial \tilde{V}}{\partial t}(\tilde{S}(t), t) + \frac{\sigma^2 \tilde{S}(t)^2}{2}\frac{\partial^2 \tilde{V}}{\partial x^2}(\tilde{S}(t), t) = 0 \quad \text{a.s.} \quad \text{for a.e. } t : \ c_F(t) \in D.$$

2 Locally bounded means that it is bounded on any bounded subset of $D \subset (0, +\infty) \times [0, T]$.

Hence

$$\frac{\partial \tilde{V}}{\partial t}(x,t) + \frac{\sigma^2 x^2}{2} \frac{\partial^2 \tilde{V}}{\partial x^2}(x,t) = 0, \quad (x,t) \in \tilde{D},$$

where $\hat{D} \triangleq \{(x,t) : \tilde{V}(x,t) > e^{-rt}F(x), \ \ t \in [0,T]\}$. This equation can be re-written as

$$\frac{\partial V}{\partial t}(x,t) + \frac{\sigma^2 x^2}{2} \frac{\partial^2 V}{\partial x^2}(x,t) = r\Big[V(x,t) - x\frac{\partial V}{\partial x}(x,t)\Big], \quad (x,t) \in D.$$

It follows that V is a solution of the boundary value problem for the parabolic equation

$$\frac{\partial V}{\partial t}(x,t) + \frac{\sigma^2 x^2}{2} \frac{\partial^2 V}{\partial x^2}(x,t) = r\Big[V(x,t) - x\frac{\partial V}{\partial x}(x,t)\Big], \quad (x,t) \in D,$$

$$V(x,T) = F(x),$$

$$V(x,t)|_{(x,t)\notin D} = F(x),$$

where

$$D = \{(x,t) : \ V(x,t) > F(x), \ \ t \in [0,T]\}. \tag{6.8}$$

Remark 6.27 In fact, (6.8) is the Bellman equation (or the dynamic programming equation) for the optimal stopping problem (6.7). It is one of the basic results of the classical theory of optimal stochastic control (for the special case of the optimal stopping problem).

Condition of smooth fitting

In fact, the solution V of (6.8) is not unique in the class of all functions that are smooth *inside D* such as was described. Therefore, not any solution of this Stefan problem represents the option fair price. Typically, the solution V of (6.8) is unique in the class of functions $V : (0 + \infty) \to \mathbf{R}$ such that the derivative $\partial V(x,t)/\partial x$ is continuous in $x \in (0, +\infty)$ (in addition to the assumptions about the properties of V *inside D* formulated above). This new assumption implies that the so-called *condition of smooth fitting* holds:

$$\frac{\partial V}{\partial x}(x,t) = \frac{dF}{dx}(x), \quad (x,t) \in \partial D. \tag{6.9}$$

The proof follows from the fact that the function $u(x,t) = V(x,t) - F(x)$ is continuous together with $\partial u(x,t)/\partial x$, and $u(x,t) \equiv 0$ for $(x,t) \notin D$.

Let us explain briefly why we need condition (6.9) to ensure that the solution V of (6.8) represents the solution of the optimal stopping problem (6.7).

It can be seen that $V(x,t) = \max(U(x,t), F(x))$, where $U(x,t)$ is a function that is absolutely continuous in $x > 0$ together with the derivative $U'_x(x,t)$ and such that $U|_D \equiv V|_D$. It follows that

$$\frac{\partial V}{\partial x}(x+0,t) - \frac{\partial V}{\partial x}(x-0,t) \geq 0 \quad \forall x > 0.$$

Let us assume that a solution V of problem (6.8) is such that

$$\frac{\partial V}{\partial x}(\hat{x}+0,t) - \frac{\partial V}{\partial x}(\hat{x}-0,t) > 0$$

for some $\hat{x}$. In this case, the process $e^{-rt}V(S(t),t)$ cannot be a supermartingale. The reason can be described, in short words, as the following: formal application by the Ito formula gives that the process $e^{-rt}V(S(t),t)$ has some 'drift coefficient' represented as the summa of a bounded process plus some unbounded positive part. The last one is generated by the derivative $\partial^2 V(x,t)/\partial x^2$ presented in the Ito formula; this derivative turns out to be a delta function at $x = \hat{x}$. (To describe the situation in a mathematically correct way, we need to use some special results from the Ito calculus related to so-called *local time*). We have proved in Corollary 6.15 that the process $\tilde{c}_F(t) = e^{-rt}c_F(t)$ is a supermartingale. Therefore, the fair price can be presented as $c_F(t) = V(S(t),t)$ only for V that is a solution of the Stefan problem (6.8) such that (6.9) holds.

Remark 6.28 We derived (6.8) only under the assumption that V is smooth enough inside D. Typically, the solvability and uniqueness of (6.8) can be proved unconditionally, i.e., under some assumptions for $F(\cdot), r, \sigma, T$ only. It is a more challenging problem.

6.7 Pricing of the American option via a binomial tree

We found that the fair price of an American option can be obtained via solution of a Stefan problem (6.8) for a parabolic equation.

The main difference with the related problem (5.24) is that the domain $D \subset \mathbf{R}^2$ (where the parabolic equation is valid) is not fixed a priori and needs to be found together with $V(\cdot)$. A boundary value problem with this feature is said to be a Stefan problem. As was mentioned above, an explicit solution of this problem is unknown even for our relatively simple case of constant (r,σ) and scalar state variable x. One possible approach is a numerical solution.

Note that Theorem 6.21 implies that

$$V(S(t\wedge\tau), t\wedge\tau) = \mathbf{E}_*\left\{e^{-r\zeta}V(S(t+\Delta)\wedge\tau), (t+\Delta)\wedge\tau)\,\middle|\,\mathcal{F}_{t\wedge\tau}\right\},$$

$$\forall t \in [0,T),\ \Delta \in [0, T-t],$$

where $\zeta \triangleq (t+\Delta)\wedge\tau - t\wedge\tau$. This leads to the following rule.

Rule 6.29 *(calculation of American option prices). For the binomial tree described in Section 6.1, the American option prices at times t_k can be estimated as $V(s_k, t_k)$, and these values can be calculated backward starting from $t_N = T$, and*

$$V(s_N, t_N) = F(S_N),$$

$$V(s_k, t_k) = \max\left[F(s_k), \tilde{V}(s_k, t_k)\right]$$

where

$$\tilde{V}(s_{k-1}, t_{k-1}) = e^{-r\Delta}\left[pV(u\,s_{k-1}, t_k) + (1-p)V(d\,s_{k-1}, t_k)\right],$$

$$k = N, N-1, \ldots, 1, \tag{6.10}$$

Here $s_k = \bar{S}(t_k)$, $t_k = kT/N$. The price at time $t_0 = 0$ is $V(s_0, 0)$, where $s_0 = S(0)$.

(Compare with Rule 6.5.) The process $\bar{S}(t_k)$ is described in Section 6.1.

The corresponding MATLAB code is given below. Note that the MATLAB Financial Toolbox has the built-in program *binprice*; it covers more options (including ones that take into account dividends) but it uses the same algorithm for the same binomial tree.

MATLAB code for pricing of the American put via a binomial tree

```
function[c]=amerput(N,s0,K,vol,T,r)
 DT=T/(N-1);
 rho=exp(-DT*r);
u=exp(vol*sqrt(DT));
 d=1/u;
s=zeros(N,N);   ff=s; s(1,1)=s0;
for k=1:N  for m=k:N
    s(k,m)=s(1,1)*u^(m-k)*d^(k-1);
end; end;
      p=(exp(DT*r)-d)/(u-d);
         for k=1:N ff(k,N)=max(0,K-s(k,N)); end;
         V=ff; for m=1:N-1 for k=1:m
         y=s(k,m); end; end;

for k=1:N-1 kk=N-k; for m=1:kk
wV=rho*(p*V(m,kk+1)+(1-p)*V(m+1,kk+1));
V(m,kk)=max(max(K-s(m,kk),0),wV); end; end;
%PutPricesTree=V
c=V(1,1);
```

Figure 6.4 shows examples of trees for European and American put prices at times t_0, t_1, t_2, t_3, for the case when the tree for the stock prices is given by Figure 6.3. It can be seen that the early exercise for American put is possible at time t_2 if $s_2 = 79.37$; if the stock price in the binomial tree model does not hit this value at time t_2, then the option is exercised at terminal time $t_3 = T$.

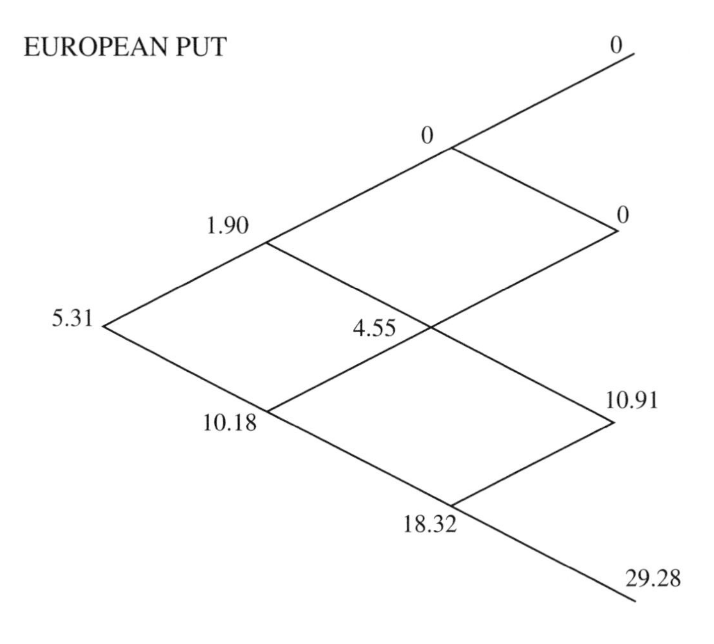

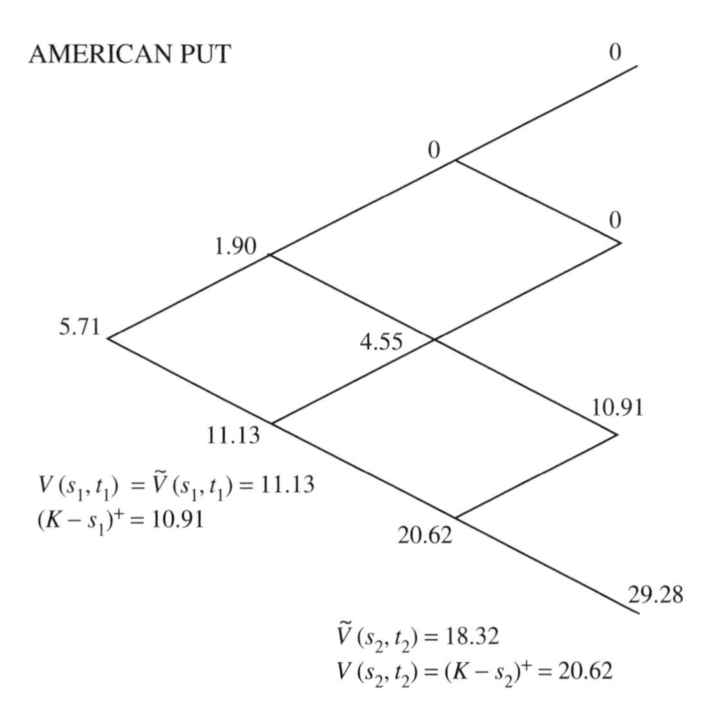

Figure 6.4 Binomial trees for European and American put.
It is assumed that $S(0) = 100$, $T = 1$, $\sigma = 0.2$, $N = 3$, $r = 0.07$, $K = 100$ (the strike price). The corresponding tree for the stock prices is given in Figure 6.3; $u = 1.12$, $d = 0.89$, $p = 0.47$.

6.8 Problems

Problem 6.30 Complete the tree in Figure 6.2.

Problem 6.31 Using a binomial tree with two time periods, calculate the price of a European put option with the strike price $K = 100$ for a three-month term with $r = 0.05$, $\sigma = 0.2$, $S(0) = 100$. (*Hint:* a three-month term corresponds to $T = 1/4$.)

Problem 6.32 Using a binomial tree with three time periods, calculate the price of an American put option with the strike price $K = 100$ for a three-month term with $r = 0.05$, $\sigma = 0.2$, $S(0) = 100$.

Problem 6.33 Let N be the number of time periods in a binomial tree. Let the prices for American and European put be calculated using a binomial tree. Is it possible that prices are different for American and European put if $N = 1$? if $N = 2$?

Challenging problem

Solve Problem 6.25.

Problem 6.34 Suggest a pricing rule for the option described in Remark 3.46. (*Hint:* use Problem 3.70.)

7 Implied and historical volatility

In this chapter, we consider again stocks and options prices in the framework of the continuous time diffusion market model. Some connections between the Black–Scholes model and the analysis of empirical market data will now be discussed.

7.1 Definitions for historical and implied volatility

Consider a risky asset (stock) with the price $S(t)$. The basic assumption for the continuous time stock price model is that the evolution of $S(t)$ is described by an Ito stochastic differential equation

$$dS(t) = a(t)S(t)dt + \sigma(t)S(t)dw(t). \tag{7.1}$$

Here $w(t)$ is a Wiener process such that $w(t) \sim N(0, t)$, i.e., the distribution law for $w(t)$ is $N(0, t)$.

Remember that the coefficient $\sigma(t)$ is said to be the *volatility*, and $a(t)$ is said to be the *appreciation rate*.

It was shown above that

$$S(T) = S(t)\exp\left(\int_t^T \mu(s)ds + \int_t^T \sigma(s)dw(s)\right), \tag{7.2}$$

where

$$\mu(t) \triangleq a(t) - \frac{\sigma(t)^2}{2}.$$

Clearly,

$$\ln S(T) = \ln S(t) + \int_t^T \mu(s)ds + \int_t^T \sigma(s)dw(s). \tag{7.3}$$

If $S(0)$, $a(t)$, and $\sigma(t)$ are deterministic, $t \in [0, T]$, then the distribution of log asset price $\ln S(T)$ is normal with mean $\ln S(0) + \int_0^T \mu(s)ds$ and variance $\int_0^T \sigma(s)^2 ds$.

Historical volatility

For the generic Black–Scholes model, we accept the hypothesis that a and σ are non-random constants. However, empirical research shows that the model may better match empirical data if the process $(a(t), \sigma(t))$ is allowed to be time-varying and random. A number of different hypotheses based on less or more sophisticated deterministic and stochastic equations for (a, σ) have been proposed in the literature. We saw that the option price depends on the volatility (but not on a), hence it is especially important for derivatives pricing to evaluate and model random volatility.

Proposition 7.1 *For any* $\varepsilon \in [0, t)$,

$$\int_{t-\varepsilon}^{t} \sigma(s)^2 ds = 2\int_{t-\varepsilon}^{t} S(s)^{-1} dS(s) - 2\ln S(t) + 2\ln S(t-\varepsilon). \tag{7.4}$$

Problem 7.2 (i) Prove Proposition 7.1. (ii) Let $\mathcal{F}_t$ be a filtration such as described in Chapter 5, and let $\tilde{S}(t) \triangleq S(t)\exp\left(\int_0^t r(s)ds\right)$, where $r(s)$ is a random bounded $\mathcal{F}_t$-adapted process. Prove that

$$\int_{t-\varepsilon}^{t} \sigma(s)^2 ds = 2\int_{t-\varepsilon}^{t} \tilde{S}(s)^{-1} d\tilde{S}(s) - 2\ln \tilde{S}(t) + 2\ln \tilde{S}(t-\varepsilon).$$

It follows that $\sigma(t)^2$ can be represented as an explicit function of past stock prices $\{S(s),\ t-\varepsilon < s < t\}$ for any $\varepsilon > 0$. This is true for the case of random and time-varying volatility, as well as for constant volatility.

In theory, formula (7.4) gives a complete description of past and present values of $\sigma(t)^2$ given past price observations for a very general class of processes $\sigma(t)$. However, this formula is not really useful in practice: the stochastic integral cannot be calculated precisely, since it must be replaced by a summa defined by the observed time series for prices. There are many working statistical methods of volatility estimation (usually, they require certain a priori assumptions about the evolution of the volatility process). We shall consider below some of these methods, including a method based on the hypothesis that volatility is constant, and a method based on so-called ARCH and GARCH models. These methods have in common that the volatility is calculated from historical stock prices. The resulting estimation is called *historical volatility*.

Note that estimation (7.4) does not give future values of volatility.

Implied volatility

Let $H_{BS,c}(x, K, \sigma, T, r)$ and $H_{BS,p}(x, K, \sigma, T, r)$ be the Black–Scholes prices at time $t = 0$ for call and put, where K is the strike price, σ is the volatility, r is the risk-free rate, T is the termination time, under the assumption that $S(0) = x$. Remember that call and put options have payoff functions $F(S(T))$, where $F(S(\cdot)) = (S(T) - K)^+$

or $F(S(\cdot)) = (K - S(T))^+$, respectively. Let us repeat the Black–Scholes formula (5.18):

$$H_{BS,c}(x, K, \sigma, T, r) = x\Phi(d_+) - Ke^{-rT}\Phi(d_-), \tag{7.5}$$

$$H_{BS,p}(x, K, \sigma, T, r) = H_{BS,c}(x) - x + Ke^{-rT},$$

where

$$\Phi(x) \stackrel{\Delta}{=} \frac{1}{\sqrt{2\pi}} \int_{-\infty}^{x} e^{-\frac{s^2}{2}} ds,$$

and where

$$d_+ \stackrel{\Delta}{=} \frac{\ln(x/K) + Tr}{\sigma\sqrt{T}} + \frac{\sigma\sqrt{T}}{2},$$

$$d_- \stackrel{\Delta}{=} d_+ - \sigma\sqrt{T}.$$

Lemma 7.3 *The Black–Scholes prices for put and call are monotonically strictly increasing functions with respect to* σ.

Proof. It suffices to find the derivatives $\partial H_{BS,c}/\partial\sigma$ and $\partial H_{BS,p}/\partial\sigma$ (i.e., to solve the problem below). Then it will be seen that these derivatives are positive. □

Problem 7.4 Find the derivatives mentioned in the proof of Lemma 7.3.

Lemma 7.3 has clear economical sense: the price for options is higher for a market with bigger uncertainty (i.e., with a bigger σ). In other words, a more *volatile* market defines a higher price for risk.

By Lemma 7.3, there is a one-to-one correspondence between the Black–Scholes option prices for put or call and the volatilities (if all other parameters are fixed).

Consider a market where a stock and options on this stock are traded. Let $V_{BS}(\sigma) = V_{BS}(S(0), K, \sigma, T, r)$ denotes the Black–Scholes price for either put or for call, where K is the strike price, σ is the volatility, and r is the risk-free rate. It is reasonable to assume that the market prices of options are meaningful and that they reflect some essential market factors. For instance, one can assume that a market price is based on offers made by major financial institutions, who set prices accurately using sophisticated models and refined methods to estimate volatility.

Assume that an investor does not collect the historical prices and does not measure volatility from the historical prices, but observes stock and option prices. This investor may try to calculate volatility by solving the equation $V_{BS}(\sigma) = \hat{V}$ with respect to σ, where $\hat{V}$ is the market price of the option. (It follows from Lemma 7.3 that $V_{BS}(\sigma)$ is a strictly increasing function in σ, so this equation is solvable.) The solution $\sigma = \sigma_{imp}$ of the equation $V_{BS}(\sigma) = \hat{V}$ is called implied volatility.

Definition 7.5 *A value σ_{imp} is said to be implied volatility at time $t = 0$ for the call option given K, r, T, if the current market price of the option at time $t = 0$ can be represented as $H_{BS,c}(S(0), K, T, \sigma_{imp}, r)$, where $H_{BS,c}(S(0), K, T, \sigma, r)$ is the Black–Scholes price for call, where K is the strike price, σ is the volatility, r is the risk-free rate, and T is the terminal time.*

The definition for the implied volatility for a put option is similar.

If a market is exactly the Black–Scholes market with constant σ, then σ_{imp} does not depend on (T, K), and it is equal to this σ (which is also the historical volatility). However, in the real market, σ_{imp} depends usually on (T, K) (and on the type of option), and the implied volatility differs from the historical volatility. In this case, we can conclude that the Black–Scholes model does not describe the real market perfectly, and its imperfections can be characterized by the gap between the historical and implied volatilities.

Varying K and T gives different patterns for implied volatility. Similarly, the evolving price $S(t)$ gives different patterns for implied volatility for different t for a given K.

The most famous pattern is the so-called *volatility smile* (or *volatility skew*) that describes dependence of σ_{imp} on K. Very often these patterns have the shape of a smile (or sometimes skew). These shapes are carefully studied in finance, and very often they are used by decision-makers. They are considered to be important market indicators, and there are some empirical rules about how to use them in option pricing. For instance, there is some empirical evidences that the Black–Scholes formula gives a better estimate of at-the-money options (i.e., when $K \sim S(t)$), and a larger error for in-the-money and out-of-the-money options (i.e., the historical volatility is closer to the implied volatility when $K \sim S(t)$). In addition, different models for evolution of random volatility are often tested by comparing the shape of volatility smiles resulting from simulation with the volatility smile obtained from real market data.

7.2 Calculation of implied volatility

By Lemma 7.3, any Black–Scholes price $V(\sigma)$ for call or put is a monotonic increasing function in σ. An example of its shape is given in Figure 7.1. The code that creates this graph is given below.

MATLAB code for representation of the Black–Scholes price as a function of volatility

```
function[v]=volprice(N,eps,s,K,T,r) v=zeros(1,N);
op=v; for k=1:N, v(k)=eps*k; op(k)=call(x,K,v(k),T,r); end;
plot(v(1:N),op(1:N),'b-');
```

Since V is monotonic, finding the root of the scalar equation $V(\sigma) = \hat{V}$ is a simple numerical problem. A very straightforward, and the simplest, solution is

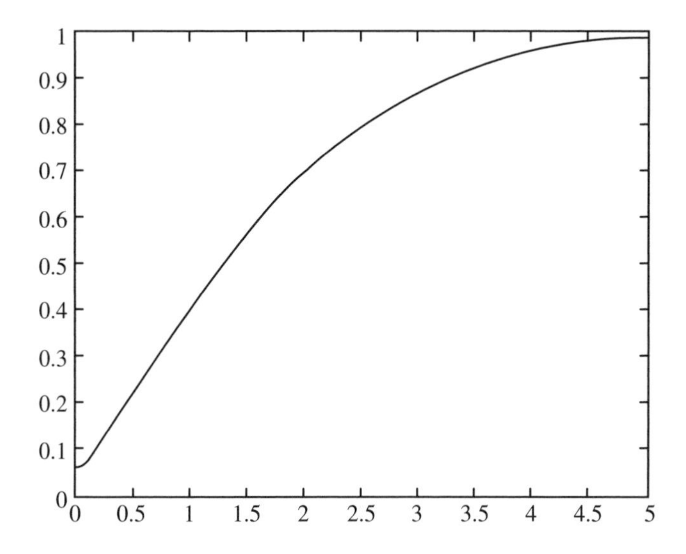

Figure 7.1 Option price as a function of volatility.
It is assumed that $(N, eps, s, K, T, r) = (500, 0.01, 1, 1, 1, 0.07)$.

to calculate all values of $V(\sigma)$ with some small enough step (i.e., to calculate the Black–Scholes price as a function of volatility), and then find the best matching value. The corresponding code is given below.

MATLAB code for extracting implied volatility from the call price

```
function[vol]=impvol(N,eps,price,s,K,T,r) v=zeros(1,N); op=v;
delta=100; m=0; for k=1:N, v(k)=eps*k; op(k)=call(x,K,v(k),T,r);
    if abs(op(k)-price)<delta
    delta=abs(op(k)-price); m=k; end;
end;
vol=v(m); pr=price*ones(1,N);
plot(v(1:N),op(1:N),'b-',v(1:N),pr(1:N),'r-.');
```

Another way is to use any modification of the gradient method (for this, the explicit formula for $V'(\sigma)$ can be used). For example, the Newton–Raphson method of solution of the equation $f(x) = 0$ is to find a root via sequence $x_{k+1} = x_k - f'(x)/f(x)$, $k = 0, 1, 2, \ldots$. Note that MATLAB Financial Toolbox has a program *blsimpv* for calculating implied volatility that is based on a precise numerical solution of the scalar equation.

Problem 7.6 Suggest an algorithm of calculation of implied volatility using a gradient method.

7.3 A simple market model with volatility smile effect

Consider a single stock market model from Example 5.63 with traded options on that stock. For this model, all options are priced on the basis of the hypothesis that, for any risk-neutral measure, volatility is random, independent of time, and can take only two values, σ_1 and σ_2, with probabilities p and $1-p$ respectively, where $p \in [0,1]$ is given. Further, assume that the pricing rule for an option with payoff function $F(S(T))$ is $e^{-rT}\mathbf{E}_* F(S(T))$. Here $\mathbf{E}_*$ is the expectation generated by the measure $\mathbf{P}_*$ such that

$$\mathbf{P}_*(A) = p\mathbf{P}_*(A \mid \sigma = \sigma_1) + (1-p)\mathbf{P}_*(A \mid \sigma = \sigma_2)$$

for all random events A. In particular, the price of the call option with the strike price K and expiration time T is

$$H_{BS,c}(S(0), K, T, \sigma_1, r) + (1-p)\, H_{BS,c}(S(0), K, T, \sigma_2, r). \tag{7.6}$$

Figure 7.2 represents the implied volatility calculated for this market using the code given below. It can be seen that even this very simple volatility model generates a volatility smile.

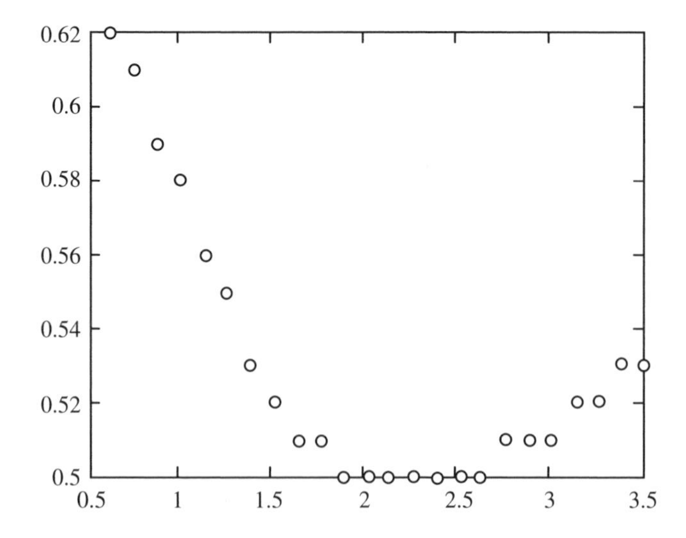

Figure 7.2 Implied volatility, the market with random volatility from Example 5.63. It is assumed that $(N, eps, s, T, r) = (24, 0.125, 2, 1, 0.1)$.

MATLAB code for modelling of volatility smile (Example 5.63)

```
function[v]=volsmile(N,eps,s,T,r) p=0.5; v1=0.3; v2=0.7; N1=100;
eps1=1/100; v=zeros(1,N);  op=v; K=v; K0=max(0.1,s-N*eps/2); for
k=1:N K(k)=K0+eps*k;
    op(k)=p*call(x,K(k),v1,T,r)+(1-p)*call(x,K(k),v2,T,r);
    v(k)=impliedvol(N1,eps1,op(k),s,K(k),T,r);
end; plot(K(1:N),v(1:N),'ro');
```

7.4 Problems

Problem 7.7 Assume that we observe prices $S_0^{(i)}$ for two different stocks at time $t = 0, i = 1, 2$. We assume that the risk-free rate is constant and known. Assume also that European call options on these two stocks, with the same strike price K and same expiration time $T > 0$, have market prices $C^{(i)}$ at time $t = 0, i = 1, 2$. Let $C^{(1)} > C^{(2)}$. Let $\sigma_{imp}^{(i)}$ be the corresponding implied volatilities, $i = 1, 2$. Indicate which statement is most correct and explain your answer:

(a) If $S_0^{(1)} < S_0^{(2)}$, then $\sigma_{imp}^{(1)} < \sigma_{imp}^{(2)}$. (b) If $S_0^{(1)} \geq S_0^{(2)}$, then $\sigma_{imp}^{(1)} \leq \sigma_{imp}^{(2)}$.

(c) If $S_0^{(1)} \leq S_0^{(2)}$, then $\sigma_{imp}^{(1)} > \sigma_{imp}^{(2)}$. (d) If $S_0^{(1)} \geq S_0^{(2)}$, then $\sigma_{imp}^{(1)} \geq \sigma_{imp}^{(2)}$.

(e) If $S_0^{(1)} \leq S_0^{(2)}$, then $\sigma_{imp}^{(1)} \geq \sigma_{imp}^{(2)}$ for some sets of parameters, and $\sigma_{imp}^{(1)} < \sigma_{imp}^{(2)}$ for some other sets of parameters.

Problem 7.8 Solve Problem 7.7 assuming put options instead of call options.

Problem 7.9 Assume that $r = 0.05$, $S(0) = 1$, and $T = 1$. Using a code that represents the Black–Scholes price of a call option as a function of volatility, draw a graph that represents the option price with this parameter as a function of the volatility. Further, assume that the price of the call option with strike price $K = 1$ is 0.25. Estimate the implied volatility using the figure.

Problem 7.10 Calculate the implied volatility using a code. Assume that the price of a call option with strike price $K = 100$ is 25, and $r = 0.05$, $S(0) = 100$, $T = 1$.

Problem 7.11 Assume that $r = 0.05$, $S(0) = 100$, $T = 0.5$. The price of a call option with strike price $K = 90$ is 25. Calculate the implied volatility.

Problem 7.12 For the market model described in Section 7.3, calculate the implied volatility for $K = S_0 \pm 0.1, \pm 0.2, \pm 0.3$ given $(S(0), r, T, p) = (1, 0.03, 1, 0.45)$.

8 Review of statistical estimation

In this chapter, we collect some core facts from mathematical statistics and statistical inference that will be used later to estimate parameters for continuous time market models.

8.1 Some basic facts about discrete time random processes

In this section, several additional definitions and facts about discrete time stochastic processes are given.

Definition 8.1 *A process ξ_t is said to be stationary (or strict-sense stationary), if the distribution of the vector $(\xi_{t_1+m}, \dots, \xi_{t_N+m})$ does not depend on m for any $N > 0, t_1, \dots, t_N$.*

Definition 8.2 *A (vector) process ξ_t is said to be wide-sense stationary, if $\mathbf{E}\xi_{t+m}$ and $\mathbf{E}\xi_{t+m}\xi_{t+\theta+m}^{\top}$ does not depend on time shift m for all t, θ, and m.*

It can happen that a process is wide-sense stationary, but it is not stationary. In fact, stationarity in a wide sense is sufficient for many applications.

Definition 8.3 *Let ξ_t, $t = 0, 1, 2, \dots$, be a discrete time random process such that ξ_t are mutually independent and have the same distribution, and $\mathbf{E}\xi_t \equiv 0$. Then the process ξ_t is said to be a discrete time white noise. If, in addition, ξ_t are normally distributed, then it is said to be a Gaussian white noise.*

The main feature of a white noise process is that its value cannot be predicted. More precisely, whatever large statistic one collects, it will not help to improve forecasting. A typical example is the outcome of the coin tossing game: it is meaningless to collect historical data to predict the result. The standard approach in data analysis is to represent the observable process as a summa of a white noise and a meaningful process that evolves under a certain law (and it appears that this meaningful part is zero for the case of the coin tossing game).

On the other hand, a white noise can be used as a basic construction block for the modelling of random processes. For instance, one can model a stationary discrete time process with given characteristics as the output of an autoregression, where a white noise is used as the input. In contrast to the white noise, the resulting output process can be statistically forecasted in a certain sense.

Sometimes in the literature white noise is defined as a wide-sense stationary process with no correlation and with zero mean.

Let x_k and y_k, $k = 1, \dots, T$, be two sequences.

Definition 8.4 *The sample mean of the sequence* $\{x_k\}$ *is* $(1/T)\sum_{k=1}^{T} x_k$. *The sample second moment of the sequence* $\{x_k\}$ *is* $(1/T)\sum_{k=1}^{T} x_k^2$. *The sample variance of* $\{x_k\}$ *is* $(1/T)\sum_{k=1}^{T}\left(x_k - (1/T)\sum_{k=1}^{T} x_k\right)^2$. *The sample covariance for sequences* $\{x_k\}$ *and* $\{y_k\}$ *is* $(1/T)\sum_{t=1}^{T} y_t x_t - (1/T)\sum_{k=1}^{T} y_k (1/T)\sum_{t=1}^{T} x_t$.

Theorem 8.5 *Let* $\{x_k\}$, $\{y_k\}$, *and* $\{(x_k, y_k)\}$ *be wide-sense stationary random processes. Then*

$$\mathbf{E}x_k = \lim_{T\to+\infty}\frac{1}{T}\sum_{k=1}^{T} x_k,\quad \mathbf{E}x_k^2 = \lim_{T\to+\infty}\frac{1}{T}\sum_{k=1}^{T} x_k^2,\quad \mathbf{E}x_k y_k = \lim_{T\to+\infty}\frac{1}{T}\sum_{k=1}^{T} x_k y_k.$$

Corollary 8.6 *Let* x_k *be i.i.d. (independent identically distributed) and have the same probability distributions as a random variable* X. *Further, let* y_k *be i.i.d., and let* (x_k, y_k) *be i.i.d. and have the same probability distributions as a random variable* Y *and a random vector* (X, Y) *respectively. Then*

$$\mathbf{E}X = \lim_{T\to+\infty}\frac{1}{T}\sum_{k=1}^{T} x_k,\quad \mathbf{E}X^2 = \lim_{T\to+\infty}\frac{1}{T}\sum_{k=1}^{T} x_k^2,\quad \mathbf{E}XY = \lim_{T\to+\infty}\frac{1}{T}\sum_{k=1}^{T} x_k y_k.$$

One may think that, under the assumptions of Corollary 8.6, x_k are random variables generated as the results of measurement of the random variable X. In this case, it is common to use the following definition.

Definition 8.7 *Assume that we have results* $x_1, \dots, x_T$ *of measurements of a random variable* X. *The sample mean of* X *is* $(1/T)\sum_{k=1}^{T} x_k$. *The sample second moment of* X *is* $(1/T)\sum_{k=1}^{T} x_k^2$. *The sample variance of* X *is* $(1/T)\sum_{k=1}^{T}\left(x_k - (1/T)\sum_{k=1}^{T} x_k\right)^2$.

A similar definition exists for sample covariance and sample correlations of two sequences of random variables, and for results of measurements of two random variables.

Remark 8.8 Note that some authors prefer to define the sample variance of $\{x_k\}$ as $1/(T-1)\sum_{k=1}^{T}\left(x_k - (1/T)\sum_{k=1}^{T} x_k\right)^2$. This is an unbiased estimate.

The distinction between two definitions may be a source of confusion, and one should be aware which definition is used. We shall use the definition with $1/T$.

For simplicity, we shall use below stationary processes and white noise in the sense of Definitions 8.1–8.3, but all results are valid for wide-sense stationary processes and for the white noise defined as a wide-sense stationary process with no correlation and zero mean.

8.2 Simplest regression and autoregression

The first-order regression model can be described by a one-dimensional equation

$$y_t = \beta_0 + \beta x_t + \varepsilon_t, \quad t = 1, 2, \dots. \tag{8.1}$$

Here y_t and x_t represent observable discrete time processes; y_t is called the regressand, or dependent variable, x_t is called the regressor, or explanatory variable, ε_t is an unobserved error term, $\beta_0 \in \mathbf{R}$ and $\beta \in \mathbf{R}$ are parameters that are usually unknown.

The standard assumption is that

$$\mathbf{E}\varepsilon_t = 0, \quad \mathbf{E}x_t\varepsilon_t = \mathrm{Cov}(x_t, \varepsilon_t) = 0 \quad \forall t. \tag{8.2}$$

Special case: autoregression (AR)

Let us describe the *first-order autoregressive process*, AR(1), as

$$y_t = \beta_0 + \beta y_{t-1} + \varepsilon_t, \tag{8.3}$$

where ε_t is a white noise process, $\beta_0, \beta \in \mathbf{R}$ are parameters.

The AR(1) model is a special case of the simplest regression (8.1), where $x_t = y_{t-1}$.

It can be shown that ε_t is uncorrelated with $\{y_s\}_{s<t}$.

If $-1 < \beta < 1$, then there exists a stationary process $\bar{y}_t$ such that $\mathbf{E}|y_t - \bar{y}_t|^2 \to 0$ as $t \to +\infty$.

If $\beta = 1$ and $\beta_0 = 0$ in (8.3), then y_t is a *random walk* (see Definition 2.6). A random walk is non-stationary and it does not converge to any stationary process.

In fact, if $|\beta| \geq 1$, then $\mathrm{Var}\, y_t \to +\infty$ as $t \to +\infty$. This implies that many standard tools for forecasting and testing coefficients etc. are invalid. To avoid this, we can try to study changes in y_t instead: for example, the differences $z_t = y_t - y_{t-1}$ may converge to a stationary process. If not, the differences $z_t - z_{t-1}$ (i.e., the second differences $y_t - 2y_{t-1} + y_{t-2}$) may converge to a stationary process.

8.3 Least squares (LS) estimation

Consider again the basic regression model (8.1)–(8.2). Suppose that either β_0 or β, or the entire pair (β_0, β) is unknown, and that we have a sample of data $\{(y_t, x_t)\}_{t=1}^{T}$.

The problem arises as to how to estimate (β_0, β) most effectively using all results of observations available.

Let $\hat{\beta}_0$ and $\hat{\beta}$ be some estimates of parameters β_0 and β.

The term

$$\hat{y}_t \triangleq \hat{\beta}_0 + \hat{\beta}x_t$$

is called the fitted value of observation. The term

$$u_t \triangleq y_t - \hat{y}_t = y_t - \hat{\beta}_0 - \hat{\beta}x_t$$

is called the residual, or fitted residual.

Definition 8.9 *The LS estimator (or OLS – ordinary least squares estimator) of β_0 and β chooses the estimates $\hat{\beta}_0$ and $\hat{\beta}$ that minimize the loss function*

$$\sum_{t=1}^{T} u_t^2. \tag{8.4}$$

Remark 8.10 LS methods allow many modifications. For instance, it can be reasonable to minimize $\sum_{t=1}^{T} c_t u_t^2$, where $c_t > 0$ are some given weight coefficients. For example, it can be useful when the variance of errors ε_t is not constant. In that case, one can take $c_t = (\mathbf{E}\varepsilon_t^2)^{-1}$: it is natural, since terms with more uncertainty should have reduced impact on the decision.

Note that least squares is only one of many possible ways to estimate regression coefficients. Some of them will also be discussed here.

Remark 8.11 Note that (β_0, β) is the true (unobservable) value which we estimate to be $(\hat{\beta}_0, \hat{\beta})$. Even if (β_0, β) is an unknown deterministic vector, $(\hat{\beta}_0, \hat{\beta})$ is a random vector since it is calculated as a function of the random sample of $\{(y_t, x_t)\}$.

We shall use below the following elementary fact.

First-order condition for minimizing for square polynomial: assume that we want to find the value of e that solves the problem

$$\text{Minimize } f(x) \text{ over } x \in \mathbf{R},$$

where $f(x) = ax^2 + bx + c$, $a > 0$. Then the minimum exists, it is unique, and it is for the value of x where $df(x)/dx = 0$.

Since $\sum_{t=1}^{T} u_t^2$ is a quadratic polynomial with respect to $\hat{\beta}_0$ and with respect to β, it follows that the minimum of this summa is achieved for $\hat{\beta}_0$ and $\hat{\beta}$ such that

$$\frac{\partial}{\partial \beta_0} \sum_{t=1}^{T} u_t^2 = 0, \quad \frac{\partial}{\partial \beta} \sum_{t=1}^{T} u_t^2 = 0. \tag{8.5}$$

Case of known β

Proposition 8.12 *Consider a model with known and given* β. *Then the LS estimate for* β_0 *is*

$$\hat{\beta}_0 \triangleq \frac{1}{T}\sum_{t=1}^{T}(y_t - \beta x_t).$$

Proof. By (8.5),

$$\begin{aligned}\frac{\partial}{\partial \hat{\beta}_0}\sum_{t=1}^{T} u_t^2 &= \frac{\partial}{\partial \hat{\beta}_0}\sum_{t=1}^{T}(y_t - \hat{\beta}_0 - \beta x_t)^2 \\ &= -2\sum_{t=1}^{T} u_t = -2\sum_{t=1}^{T}(y_t - \hat{\beta}_0 - \beta x_t) = 0.\end{aligned} \tag{8.6}$$

The solution is straightforward. By (8.6), we have $T\hat{\beta}_0 = \sum_{t=1}^{T}(y_t - \beta x_t)$. Hence $\hat{\beta}_0 = (1/T)\sum_{k=1}^{T}(y_t - \beta x_t)$. This completes the proof. □

Case of known $\beta_0 = 0$

Proposition 8.13 *Consider a model with known* $\beta_0 = 0$. *In addition, we assume that* y_t *and* x_t *have zero expectation. Then the LS estimate for* β *is*

$$\hat{\beta} = \frac{(1/T)\sum_{t=1}^{T} y_t x_t}{(1/T)\sum_{t=1}^{T} x_t^2}. \tag{8.7}$$

Proof. By (8.5),

$$\begin{aligned}\frac{\partial}{\partial \hat{\beta}}\sum_{t=1}^{T} u_t^2 = \frac{\partial}{\partial \hat{\beta}}\sum_{t=1}^{T}(y_t - \hat{\beta} x_t)^2 &= -2\sum_{t=1}^{T}(y_t - \hat{\beta} x_t)x_t \\ &= -2\sum_{t=1}^{T} u_t x_t = 0.\end{aligned} \tag{8.8}$$

Then

$$\sum_{t=1}^{T}(y_t - \hat{\beta} x_t)x_t = \sum_{t=1}^{T}(y_t x_t - \hat{\beta} x_t^2) = \sum_{t=1}^{T} y_t x_t - \hat{\beta}\sum_{t=1}^{T} x_t^2 = 0.$$

Then (8.7) follows. This completes the proof. □

Case of unknown (β_0, β)

Theorem 8.14 *For the case of unknown* (β_0, β)*, the LS estimator is*

$$\hat{\beta}_0 = \frac{1}{T}\sum_{k=1}^{T} y_k - \hat{\beta}\frac{1}{T}\sum_{k=1}^{T} x_k, \tag{8.9}$$

$$\hat{\beta} = \frac{(1/T)\sum_{t=1}^{T} y_t x_t - (1/T)\sum_{k=1}^{T} y_k (1/T)\sum_{t=1}^{T} x_t}{(1/T)\sum_{t=1}^{T} x_t^2 - (1/T)\sum_{k=1}^{T} x_k (1/T)\sum_{t=1}^{T} x_t}. \tag{8.10}$$

To make these formulae more visible, we can rewrite them as

$$\hat{\beta}_0 = \frac{1}{T}[(y, \mathbf{1}) - \hat{\beta}(x, \mathbf{1})], \tag{8.11}$$

$$\hat{\beta} = \frac{(y, x) - (y, \mathbf{1})(x, \mathbf{1})/T}{(x, x) - (x, \mathbf{1})^2/T}. \tag{8.12}$$

Here $\mathbf{1} \triangleq (1, \dots, 1)^\top \in \mathbf{R}^T$, $x = (x_1, \dots, x_T)^\top \in \mathbf{R}^T$, $y = (y_1, \dots, y_T)^\top \in \mathbf{R}^T$, and $(x, y) = \sum_i x_i y_i$.

Remark 8.15 Remember that $(1/T)\sum_{t=1}^{T} y_t x_t - (1/T)\sum_{k=1}^{T} y_k (1/T)\sum_{t=1}^{T} x_t$ is the sample covariance. It follows that the estimator for β is the sample covariance of y_t and x_t, divided by the sample covariance of the regressor x_t.

Remark 8.16 If (x_t, y_t) is a stationary process, then there is a convergence of LS estimates:

$$\hat{\beta}_0 \to \mathbf{E}y_t - \hat{\beta}\mathbf{E}x_t, \quad \hat{\beta} \to \frac{\mathbf{E}y_t x_t - \mathbf{E}y_t \mathbf{E}x_t}{\mathbf{E}x_t^2 - (\mathbf{E}x_t)^2} = \frac{\mathrm{Cov}(y_t, x_t)}{\mathrm{Var}\, x_t^2} \quad \text{as } T \to +\infty.$$

Proof of Theorem 8.14. By (8.5),

$$\frac{\partial}{\partial \hat{\beta}_0}\sum_{t=1}^{T}(y_t - \hat{\beta}_0 - \hat{\beta}x_t)^2 = -2\sum_{t=1}^{T}(y_t - \hat{\beta}_0 - \hat{\beta}x_t) = 0, \tag{8.13}$$

$$\frac{\partial}{\partial \hat{\beta}}\sum_{t=1}^{T}(y_t - \hat{\beta}_0 - \hat{\beta}x_t)^2 = -2\sum_{t=1}^{T}(y_t - \hat{\beta}_0 - \hat{\beta}x_t)x_t = 0. \tag{8.14}$$

We have two equations and two unknowns ($\hat{\beta}_0$ and $\hat{\beta}$). The solution is straightforward. By (8.13), we have

$$T\hat{\beta}_0 = \sum_{t=1}^{T}(y_t - \hat{\beta}x_t).$$

Hence

$$\hat{\beta}_0 = \frac{1}{T}\sum_{k=1}^{T} y_k - \hat{\beta}\frac{1}{T}\sum_{k=1}^{T} x_k.$$

Substituting into (8.14) gives

$$\sum_{t=1}^{T}\left[y_t - \frac{1}{T}\sum_{k=1}^{T} y_k + \hat{\beta}\frac{1}{T}\sum_{k=1}^{T} x_k - \hat{\beta}x_t\right]x_t$$

$$= \sum_{t=1}^{T}\left(y_t x_t - \frac{1}{T}\sum_{k=1}^{T} y_k x_t + \hat{\beta}\frac{1}{T}\sum_{k=1}^{T} x_k x_t - \hat{\beta}x_t^2\right) = 0.$$

Hence

$$\hat{\beta} = \frac{\sum_{t=1}^{T}(y_t x_t - (1/T)\sum_{k=1}^{T} y_k x_t)}{\sum_{t=1}^{T}(x_t^2 - (1/T)\sum_{k=1}^{T} x_k x_t)} = \frac{\sum_{t=1}^{T} y_t x_t - (1/T)\sum_{k=1}^{T} y_k \sum_{t=1}^{T} x_t}{\sum_{t=1}^{T} x_t^2 - (1/T)\sum_{k=1}^{T} x_k \sum_{t=1}^{T} x_t}.$$

Then (8.9) and (8.10) follow. This completes the proof of Theorem 8.14. □

Remark 8.17 Note that (8.13) and (8.14) can be rewritten as

$$\frac{1}{T}\sum_{t=1}^{T} u_t = 0, \qquad \frac{1}{T}\sum_{t=1}^{T} u_t x_t = 0.$$

This can be seen as the sample analogues of the assumptions in (8.2) that $\mathbf{E}\varepsilon_t = 0$ and $\mathrm{Cov}(x_t, \varepsilon_t) = 0$.

Problem 8.18 Estimate parameters (β_0, β) for the series

(i) $x = (1, -2, 3, 1, 0, 0, 1)$; $y = (1, -2, 3, 1, 0.5, 0, 1)$;
(ii) $y_t = \ln S_1(t) - \ln S_1(t-1)$, $x_t = \ln S_2(t) - \ln S_2(t-1)$, where $S_1(t)$ and $S_2(t)$ are daily prices stored in some files.

Problem 8.18 may be solved using the following codes.

MATLAB code for the LS estimator for Problem 8.18 (i)

```
function[b0,b]=ols(m)
m=7;
x=[1 -2 3 1 0 0 1]; y=[1 -2 3 1 0.5 0 1];
%load OLSdat -ascii; xxx=OLSdat; Another way to load data
one=ones(size(x)); b=1;
%figure(1);plot(t(1:m1),xxx(1:m1),'b-');
```

```
b=(sum(y .*x)-sum(y .*one)*sum(x .*one)/m)/(sum(x .*x)
-sum(x.*one)^2/m);
b0=sum(y .*one)/m-b*sum(x .*one)/m;
```

MATLAB code for the LS estimator for Problem 8.18 (ii)

It is assumed that the prices are stored in the ASCII files NCP.mat and ANZ.mat placed in the working directory; these files contain matrices, where the second column represents the prices under consideration.

```
function[b0,b]=ols(m)
load ncp -ascii; xxx=ncp;
load anz -ascii;
yyy=anz; yy=ones(m,1); xx=yy; m1=m-1; y=ones(m1,1); x=y; one=y; b=1;
t=one; for k=1:m t(k)=k;
    yy(k)=log(yyy(k,2)); xx(k)=log(xxx(k,2));
end; for k=1:m1
    y(k)=(yy(k+1))-(yy(k)); x(k)=(xx(k+1))-(xx(k));
end;  plot(t(1:m1),x(1:m1)-y(1:m1),'b-');

b=(sum(y .*x)-sum(y .*one)*sum(x .*one)/m1)/(sum(x .*x)
-sum(x.*one)^2/m1); b0=sum(y .*one)/m1-b*sum(x .*one)/m1;
```

8.4 The LS estimate of the variance of the error term

Consider the linear regression model (8.1).

Definition 8.19 *The value*

$$\hat{\sigma}^2 = \frac{1}{T-1}\sum_{t=1}^{T} u_t^2 \tag{8.15}$$

is said to be the LS estimate of the variance of the error term.

Let $\mathcal{F}_t$ be the filtration generated by the observations, i.e., by the process (x_t, y_t).

Definition 8.20 *(homoscedasticity and heteroscedasticity). We say that the errors ε_t are homoscedastic if there exists a constant σ^2 such that $\mathbf{E}\{\varepsilon_t^2|\mathcal{F}_{t-1}\} = \sigma^2 > 0$ for all t. Otherwise, the errors are said to be heteroscedastic.*

Theorem 8.21 *Let the errors be homoscedastic, and let $\mathbf{E}\{\varepsilon_t^2 \mid \mathcal{F}_{t-1}\} = \sigma^2 > 0$ for all t. Then, under some additional conditions, the estimate (8.15) of σ^2 is unbiased.*

8.5 The case of AR(1)

The AR(1) model is a special case of the simplest regression (8.1), where $x_t = y_{t-1}$ (since ε_t and y_{t-1} are uncorrelated). Hence AR(1) can be estimated with LS.

Remark 8.22 For LS applied for the AR(1) model and series $(y_0,\dots,y_T)$, we use the initial value y_0 only to produce $x_1 = y_0$; we apply the LS for $\{(y_t,x_t)\}_{t=1}^T = \{(y_t,y_{t-1})\}_{t=1}^T$.

Problem 8.23 Consider the sequence $(1,-1,0.5)$. Estimate the parameters for the autoregression model

$$y_k = \beta y_{k-1} + \varepsilon_k, \qquad k = 1,2,\dots,$$

where ε_k is a white noise process, $\mathbf{E}\varepsilon_t^2 = \sigma^2$, and where a is an unknown parameter.

Solution. This model is a special case of the regression model

$$y_k = \beta x_k + \varepsilon_k, \quad \text{where} \quad (x_1,x_2) = (1,-1), \quad (y_1,y_2) = (-1,0.5).$$

The LS estimator (the case of $\beta_0 = 0$) estimates β as

$$\hat{\beta} = \frac{\sum_{i=1}^T x_i y_i}{\sum_{i=1}^T x_i^2} = \frac{-1-0.5}{1+1} = -\frac{3}{4}, \qquad T = 2. \quad \square$$

8.6 Maximum likelihood

If ε_t in (8.1) are i.i.d. (i.e., independent identically distributed) with the distribution $N(0,\sigma^2)$, then (β_0,β,σ) can be estimated via maximization of a likelihood function. The probability density function of ε_t is

$$p_{\varepsilon_t}(e) = \frac{1}{\sqrt{2\pi\sigma^2}} \exp\left(-\frac{e^2}{2\sigma^2}\right). \tag{8.16}$$

Since the errors are independent, the joint probability density function of the $(\varepsilon_1,\varepsilon_2,\dots,\varepsilon_T)$ is the product of the probability density functions of each of the errors

$$p_{\varepsilon_1,\dots,\varepsilon_T}(e_1,\dots,e_T) = \prod_{t=1}^T p_{u_t}(e_t) = \left[\frac{1}{\sqrt{2\pi\sigma^2}}\right]^T \exp\left(-\sum_{t=1}^T \frac{e^2}{2\sigma^2}\right). \tag{8.17}$$

The maximum likelihood estimation method requires finding the value of $(\hat{\beta}_0,\hat{\beta},\hat{\sigma})$ that maximizes $p_{\varepsilon_1,\dots,\varepsilon_T}(u_1,\dots,u_T)$ given the observations $(x_1,\dots,x_T,y_1,\dots,y_T)$, where $u_t \stackrel{\Delta}{=} y_t - \hat{\beta}_0 - x_t\hat{\beta}$.

It is more convenient to maximize the logarithm of the probability density function, or the log-likelihood function

$$L \triangleq \ln p_{\varepsilon_1,\ldots,\varepsilon_T}(u_1,\ldots,u_T) = -\frac{T}{2}\ln(2\pi) - \frac{T}{2}\ln\hat{\sigma}^2 - \frac{1}{2\hat{\sigma}^2}\sum_{t=1}^{T}(y_t - \hat{\beta}_0 - x_t\hat{\beta})^2. \tag{8.18}$$

This likelihood function L is maximized by minimizing the last term, which is proportional to the sum of squared errors similarly to (8.4).

Corollary 8.24 *The ML estimate for (β_0, β) is the same as the LS estimate when the errors are i.i.d. with normal law (but only then).*

Remark 8.25 If errors ε_t are independent with law $N(0, \sigma_t^2)$, where $\sigma_t^2 > 0$ are known deterministic values (not necessarily equal), then the ML estimate leads to minimization of $\sum_{t=1}^{T}(1/2\hat{\sigma}^2)u_t^2$, which can be considered as a modification of the LS method with weighted terms.

Further,

$$\frac{\partial L}{\partial(\hat{\sigma}^2)} = -\frac{T}{2\hat{\sigma}^2} + \frac{T}{2\hat{\sigma}^4}\sum_{t=1}^{T}u_t^2,$$

where $u_t = y_t - \beta_0 - x_t\hat{\beta}$. Equality $\partial L/\partial(\hat{\sigma}^2) = 0$ gives the following corollary.

Corollary 8.26 *The ML estimate of σ^2 is*

$$\hat{\sigma}^2 \triangleq \frac{1}{T}\sum_{t=1}^{T}u_t^2. \tag{8.19}$$

It follows that ML and LS estimates for σ are different, but they converge as $T \to +\infty$ (see Definition 8.19). By Theorem 8.21, it follows that the ML estimate for σ is biased.

Since the ML method is based on the hypothesis that the fitted errors are Gaussian, it may require some preliminary analysis of the distributions.

8.7 Hypothesis testing

The distribution of $\hat{\beta}$ and properties of error

Repeat that the estimated coefficients are different from the true coefficients (β_0, β) in (8.1). Moreover, the estimated coefficients $(\hat{\beta}, \hat{\beta}_0)$ are random variables since they depend on a particular sample.

Let us consider the simple case when $\beta_0 = 0$, $\mathbf{E}x_t = 0$, and $\mathbf{E}y_t = 0$.[1] By (8.7),

$$\hat{\beta} = \frac{(1/T)\sum_{t=1}^{T} x_t y_t}{(1/T)\sum_{t=1}^{T} x_t^2} = \frac{(1/T)\sum_{t=1}^{T} x_t(\beta x_t + \varepsilon_t)}{(1/T)\sum_{t=1}^{T} x_t^2} = \beta + E, \tag{8.20}$$

$$E \triangleq \frac{(1/T)\sum_{t=1}^{T} x_t \varepsilon_t}{(1/T)\sum_{t=1}^{T} x_t^2}. \tag{8.21}$$

Here E is the error of estimation, and $y_t = \beta x_t + \varepsilon_t$. Remember that $\mathrm{Cov}(x_t, \varepsilon_t) \equiv 0$. Under some conditions of stationarity of (x_t, y_t), we have that

$$\frac{1}{T}\sum_{t=1}^{T} x_t^2 \to \mathbf{E}x_t^2 = \mathrm{Var}\,x_t, \quad \frac{1}{T}\sum_{t=1}^{T} x_t\varepsilon_t \to \mathrm{Cov}(x_t, \varepsilon_t) = 0 \quad \text{as } T \to +\infty. \tag{8.22}$$

In this case, $E \to 0$ as $T \to +\infty$, and we say that the estimator of β is *consistent*.

Further, it can be seen that it is difficult to find explicitly the distribution of E, even if the distribution of ε_t are known. A possible way is to find the distribution of $\hat{\beta}$ via Monte Carlo simulation. Another way is to use the asymptotic distribution for $T \to +\infty$ as an approximation.

Asymptotic distribution

Suppose that $\hat{\beta}$ is consistent, i.e., there is a non-random limit β such that $\hat{\beta} \to \beta$ as $T \to +\infty$. In this case, the distribution of $\sqrt{T}(\hat{\beta} - \beta)$ will typically converge to a non-trivial normal distribution. To see why, note that (8.20) implies

$$\sqrt{T}(\hat{\beta} - \beta) = \left(\frac{1}{T}\sum_{t=1}^{T} x_t^2\right)^{-1} \sqrt{T}\frac{1}{T}\sum_{t=1}^{T} x_t\varepsilon_t. \tag{8.23}$$

A central limit theorem ensures (under some standard conditions) that the distribution of $\sqrt{T}(1/T)\sum_{t=1}^{T} x_t\varepsilon_t$ converges to the normal distribution $N(0, v^2)$ as $T \to +\infty$, where $v^2 = \lim_{T\to+\infty} \mathrm{Var}\,\sqrt{T}(1/T)\sum_{t=1}^{T} x_t\varepsilon_t$. In other words, $\sqrt{T}(\hat{\beta} - \beta)$ has an asymptotic normal distribution, i.e.,

$$\sqrt{T}(\hat{\beta} - \beta) \to N[0, \mathrm{Var}\,(x)^{-2}v^2], \tag{8.24}$$

1 The case with $\beta \neq 0$ can also be reduced to this case in a multi-dimensional setting (see Problem 8.28).

in distributions, as $T \to +\infty$. If ε_t is independent from x_t, and ε_t are i.i.d., then (8.24) can be simplified as

$$\sqrt{T}(\hat{\beta} - \beta) \to N[0, \operatorname{Var}(\varepsilon_t) \operatorname{Var}(x_t)^{-1}]. \tag{8.25}$$

Testing hypotheses about β

Suppose that the asymptotic distribution of an estimator $\hat{\beta}$ is normal with mean β and variance v^2, i.e.,

$$\sqrt{T}(\hat{\beta} - \beta) \to N(0, v^2). \tag{8.26}$$

Let β_H be a hypothesis about the value of β. The problem of testing the hypotheses that $\beta = \beta_H$ now arises.

Confidence interval

Recall that if $\eta \sim N(0, 1)$, then

$$\mathbf{P}(|\eta| < x) = \Phi(x) - \Phi(-x).$$

Further, if $\xi \sim N(0, v^2)$ then $\xi/v \sim N(0, 1)$ and

$$\mathbf{P}(|\xi/v| < x) = \Phi(x) - \Phi(-x).$$

Let us consider the inverse problem: *given a probability $p \in (0, 1)$, find an interval that contains a random variable ξ with this probability, if $\xi \sim N(0, v^2)$.* In other words, we need to find y such that $p = \mathbf{P}(\xi \in (-y, y))$. The solution is the following: first find x such that $\Phi(x) - \Phi(-x) = p$. Further,

$$\mathbf{P}(|\xi| < v \cdot x) = \mathbf{P}(|\xi/v| < x) = \Phi(x) - \Phi(-x) = p.$$

(We assume that $v > 0$.) Hence $y = v \cdot x$.

For instance,

$$\mathbf{P}(|\xi| < 1.65v) = \mathbf{P}(|\xi/v| < 1.65) = \Phi(1.65) - \Phi(-1.65) = 0.9,$$
$$\mathbf{P}(|\xi| < 2v) = \mathbf{P}(|\xi/v| < 2) = \Phi(2) - \Phi(-2) = 0.95.$$

Hence

$$\mathbf{P}\,(|\xi| < 1.65 \cdot v) = 0.9, \quad \mathbf{P}\,(|\xi| < 2 \cdot v) = 0.95.$$

We say that $[-1.65v, 1.65v]$ is the confidence interval of significance level 10%, and that $[-2v, 2v]$ is the confidence interval of significance level 5%.

For instance, consider the hypothesis that the true value of β in the linear regression is $\beta_H = 0$. If this is true, then

$$\mathbf{P}(|\sqrt{T}(\hat{\beta} - \beta_H)/v| > 2) = \mathbf{P}(|\sqrt{T}\hat{\beta}/v| > 2) \sim 0.05.$$

We then say that we reject the hypothesis that $\beta = 0$ at the 5% significance level (95% confidence level) if the test statistic $|\sqrt{T}\hat{\beta}/v|$ is larger than 2. This means that if the hypothesis is true ($\hat{\beta} = 0$), then this rule gives the wrong decision in 5% of the cases. In fact, this estimate is only an approximation, since we are using the normal distribution as an approximation, instead of the true (and typically unknown) distribution.

Algorithm of hypothesis test

Assume that we have regression with $\beta_0 = 0$. We need to test the hypothesis that $\beta = \beta_H$ with confidence level p using an LS estimator. The discussion given above can be summarized as the following algorithm:

- Find $x > 0$ such that $\mathbf{P}(|\xi| < x) = p$ for $\xi \sim N(0,1)$, i.e.,

$$p = \mathbf{P}(|\xi| < x) = \frac{1}{\sqrt{2\pi}} \int_{-x}^{x} e^{-\frac{x^2}{2}} dx = \Phi(x) - \Phi(-x);$$

- Calculate $\hat{\beta}$ using the LS estimator;
- Calculate

$$v^2 \sim \frac{\operatorname{Var} u_t}{\operatorname{Var} x_t} = \frac{\mathbf{E}u_t^2}{\mathbf{E}x_t^2} \quad \text{as} \quad v^2 = \frac{(u,u)}{(x,x)} = \frac{(1/T)\sum_{t=1}^{T} u_t^2}{(1/T)\sum_{t=1}^{T} x_t^2},$$

  where $u_t \triangleq \hat{\beta}x_t - y_t$;
- Find $M \triangleq |\sqrt{T}(\hat{\beta} - \beta_H)/v|$;
- If $M > x$, then reject the hypothesis that $\beta = \beta_H$.

Note that the expression for v contains biased sample estimates of variances.

The following MATLAB codes support this algorithm.

MATLAB code for the inverse of $\Phi(x) - \Phi(-x)$

```
function[x]=interval(p,N) v=zeros(1,N);
eps=4/N; v=99;
for k=1:N
v(k)=eps*k;
    if abs(Phi(v(k))-Phi(-v(k))-p)<v x=v(k); end;
end;
pp=Phi(x)-Phi(-x)
```

MATLAB code for the hypothesis test

It is assumed again that the prices are stored in the ASCII files NCP.mat and ANZ.mat placed in the working directory; these files contain matrices, where the second column represents the prices under consideration. To verify that $\beta_0 \sim 0$ and $\mathbf{E}x_t \sim 0$, $\mathbf{E}y_t \sim 0$, one can find (β_0, β) and sample means preliminary with the LS estimator (see Problem 8.18 (ii)).

```
 function[b,H]=htest(p,B,m)
% p is the probability, B is the value tested, m is the number
% of prices
load ncp -ascii; xxx=ncp; load anz -ascii; yyy=anz;

yy=ones(m,1); xx=yy; m1=m-1; y=zeros(m1,1); x=y;
 one=y; b=1; t=one;
                                      for k=1:m
t(k)=k; yy(k)=log(yyy(k,2)); xx(k)=log(xxx(k,2));
                                      end;
b=sum(y .*x)/(sum(x.*x);
                                   for k=1:m1
    y(k)=(yy(k+1))-(yy(k)); x(k)=(xx(k+1))-(xx(k));
                                      end;

u=b*x-y; sigmaLS=sqrt(sum(u .*u)/(m1-1));
sigmaML=sqrt(sum(u.*u)/(m1))
v2=sum(u .*u) /sum(x.*x); v=sqrt(v2);
M=abs(sqrt(m1)*(b-B)/v); x=interval(p,1000); H=1; if M>x H=0; end;
```

8.8 LS estimate for multiple regression

The previous results can be extended for a multiple regression.

Consider the linear model

$$y_t = x_{1t}\beta_1 + x_{2t}\beta_2 + \cdots + x_{kt}\beta_k + \varepsilon_t = x_t^{\top}\beta + \varepsilon_t, \tag{8.27}$$

where y_t and ε_t are scalars, $x_t \in \mathbf{R}^k$, and $\beta \in \mathbf{R}^k$ is a vector of the true coefficients.

Let $\hat{\beta}$ be an estimate of β. Let $u_t \triangleq y_t - x_t^{\top}\hat{\beta}$ be the fitted residuals. The LS estimator minimizes the summa

$$\sum_{t=1}^{T} u_t^2 = \sum_{t=1}^{T} (y_t - x_t^{\top}\hat{\beta})^2, \tag{8.28}$$

by choosing the vector $\hat{\beta}$. The first-order conditions are

$$\sum_{t=1}^{T} x_t(y_t - x_t^{\top}\hat{\beta}) = 0_{\mathbf{R}^k}, \tag{8.29}$$

where $0_{\mathbf{R}^k}$ denotes a zero vector in $\mathbf{R}^k$, or

$$\sum_{t=1}^{T} x_t y_t = \sum_{t=1}^{T} x_t x_t^{\top} \hat{\beta},$$

which can be solved as

$$\hat{\beta} = \left(\sum_{t=1}^{T} x_t x_t^{\top} \right)^{-1} \sum_{t=1}^{T} x_t y_t. \tag{8.30}$$

Definition 8.27 *The value*

$$\hat{\sigma}^2 = \frac{1}{T-k} \sum_{t=1}^{T} u_t^2 \tag{8.31}$$

is said to be the LS estimate of the variance of the error term for linear regression model (8.27).

Problem 8.28 Show that regression (8.1) with $\beta_0 \neq 0$ is a special case of multiple regression (8.27), where $k = 2$ and $x_{1t} \equiv 1$.

Error of LS estimator for multiple regression

By (8.27)–(8.30), it follows that

$$\hat{\beta} = \left(\sum_{t=1}^{T} x_t x_t^{\top} \right)^{-1} \sum_{t=1}^{T} x_t (x_t^{\top} \beta + \varepsilon_t).$$

Hence

$$\sqrt{T}(\hat{\beta} - \beta) = \left(\frac{1}{T} \sum_{t=1}^{T} x_t x_t^{\top} \right)^{-1} \sqrt{T} \frac{1}{T} \sum_{t=1}^{T} x_t \varepsilon_t. \tag{8.32}$$

Similarly to (8.24), the distribution of this vector converges a vector normal distribution under some conditions of stationarity.

More general models

The pth-order autoregressive process, AR(p), is a straightforward extension of the AR(1) model:

$$y_t = \beta_0 + \beta_1 y_{t-1} + \beta_2 y_{t-2} + \cdots + \beta_p y_{t-p} + \varepsilon_t. \tag{8.33}$$

Here $\beta_k \in \mathbf{R}$ are parameters. It is usually assumed that ε_t does not depend on $\{y_{t-k}\}_{k=1}^{p}$.

In fact, the AR(p) model is a special case of the multiple regression (8.27), where $x_{kt} = y_{t-k}$. Hence AR(p) can be estimated with the multi-dimensional version of the LS estimator.

A qth-order moving average (MA) process is

$$y_t = \varepsilon_t + \theta_1 \varepsilon_{t-1} + \cdots + \theta_q \varepsilon_{t-q}, \tag{8.34}$$

where ε_t is a white noise.

Autoregressive-moving average models (ARMA(p,q)) are combinations of a moving average and an autoregression model. For instance, the ARMA(2,1) model has the following evolution law:

$$y_t = \beta_1 y_{t-1} + \beta_2 y_{t-2} + \varepsilon_t + \theta_1 \varepsilon_{t-1},$$

where ε_t is a white noise.

8.9 Forecasting

Conditional expectation as the best estimation

Suppose we observed first the t terms of time series $y_0, y_1, \ldots$, and we want to estimate a (random variable) ξ that can be correlated with the series. That means that we want to find a deterministic function $F_t : \mathbf{R}^{t+1} \to \mathbf{R}$ such that the estimate can be obtained as $\hat{\xi} = F_t(y_0, y_1, \ldots, y_t)$. Note that the mapping $F_t(\cdot)$ cannot depend on realizations of y_t, but can (and must) be constructed using our knowledge about the distributions of $(\xi, \{y_t\})$.

Let $\mathbf{E}_t$ denote the conditional expectation given $(y_0, y_1, \ldots, y_t)$, i.e.,

$$\mathbf{E}_t\{\,\cdot\,\} \triangleq \mathbf{E}\{\,\cdot\,|\,y_0, \ldots, y_t\}.$$

In other words, it is the conditional expectation with respect to the σ-algebra generated by the random vector $(y_0, \ldots, y_t)$. (See Definitions 1.42–1.44.) Let $\mathrm{Var}_t\xi = \mathbf{E}_t(\xi^2 - (\mathbf{E}_t\xi)^2)$ (i.e., it is the conditional variance, or the variance for the conditional probability space under the condition that $(y_0, \ldots, y_t)$ is known).

Let $\mathbf{E}\xi^2 < +\infty$, let $\hat{\xi} \triangleq \mathbf{E}_t\xi$, and let η be any random variable that is measurable with respect to $(y_0, \ldots, y_t)$ and such that $\mathbf{E}\eta^2 < +\infty$. By Definition 1.36 for the conditional expectation, it follows that $\mathbf{E}(\xi - \hat{\xi})^2 \le \mathbf{E}(\xi - \eta)^2$. In that sense, $\hat{\xi} = \mathbf{E}_t\xi$ is the best estimate of ξ.

Repeat that $\hat{\xi} = \mathbf{E}_t\xi$ can be represented as $F_t(y_0,\ldots,y_t)$ for some function $F_t : \mathbf{R}^t \to \mathbf{R}$ (see Theorem 1.45).

Forecasting for AR(1)

Forecasting with known parameters

Suppose we have estimated an AR(1) described by (8.3), i.e.,

$$y_{t+1} = \beta_0 + \beta y_t + \varepsilon_{t+1} \tag{8.35}$$

with white noise ε_t such that $\sigma^2 \triangleq \mathrm{Var}\,(\varepsilon_t) = \text{const}$. Assume that we know β_0, β and σ^2 (for instance, we had estimated them by the LS method precisely enough).

We want to forecast y_{t+1} using information available in t. In fact, $\mathbf{E}_t y_k$ is the best forecast of y_k, $k > t$ (in the sense of Definition 1.36). Since $\mathbf{E}_t\varepsilon_{t+1} = 0$, then

$$\mathbf{E}_t y_{t+1} = \beta y_t + \beta_0. \tag{8.36}$$

Further,

$$\begin{aligned} y_{t+2} &= \beta y_{t+1} + \beta_0 + \varepsilon_{t+2} = \beta(\beta y_t + \beta_0 + \varepsilon_{t+1}) + \beta_0 + \varepsilon_{t+2} \\ &= \beta^2 y_t + \beta\beta_0 + \beta\varepsilon_{t+1} + \beta_0 + \varepsilon_{t+2}. \end{aligned}$$

Since $\mathbf{E}_t\varepsilon_{t+1} = 0$ and $\mathbf{E}_t\varepsilon_{t+2} = 0$, it follows that

$$\mathbf{E}_t y_{t+2} = \beta^2 y_t + \beta\beta_0 + \beta_0. \tag{8.37}$$

Repeating this, we obtain

$$\mathbf{E}_t y_{t+s} = \beta^s y_t + [\beta^{s-1} + \beta^{s-2} + \cdots + 1]\beta_0 = \beta^s y_t + \frac{1-\beta^s}{1-\beta}\beta_0 \quad s = 1,2,\ldots. \tag{8.38}$$

(The last equality holds if $\beta \neq 1$.)

Remark 8.29 If $\beta = \beta_0 = 0$ in (8.3), then y_t is a white noise process. By (8.38), the forecast in this case is a constant (zero) for all forecasting horizons.

Conditional variance of the forecast error

We have that

$$\begin{aligned} &y_{t+1} - \mathbf{E}_t y_{t+1} = \varepsilon_{t+1}, && \mathrm{Var}_t y_{t+1} = \mathrm{Var}_t \varepsilon_{t+1} = \mathrm{Var}\,\varepsilon_{t+1} = \sigma^2, \\ &y_{t+2} - \mathbf{E}_t y_{t+2} = \beta\varepsilon_{t+1} + \varepsilon_{t+2}, && \mathrm{Var}_t y_{t+2} = \beta^2\sigma^2 + \sigma^2. \end{aligned}$$

Similarly, it can be shown that

$$v_s \triangleq \mathbf{E}_t(y_{t+s} - \mathbf{E}_t y_{t+s})^2 = [1 + \beta^2 + \beta^4 + \cdots + \beta^{2(s-1)}]\sigma^2 = \frac{1-\beta^{2s}}{1-\beta^2}\sigma^2,$$

where the last equality holds if $|a| \neq 1$. In fact, $v_s = \mathrm{Var}_t y_{t+s}$, i.e., it is the conditional variance given $y_1, \dots, y_t$ (the variance on the conditional probability space).

Problem 8.30 Let

$$y_k = \beta y_{k-1} + \varepsilon_k, \qquad k = 1, 2, \dots,$$

where ε_k is the white noise process, $\mathbf{E}\varepsilon_t^2 = \sigma^2$, and where $\beta = 2, \sigma^2 = 1/2$. Find $\mathbf{E}\{y_9|y_1, \dots, y_6\}$ and $\mathbf{E}\{y_9^2|y_1, \dots, y_6\}$.

Solution. We have

$$\begin{aligned} y_9 = \beta y_8 + \varepsilon_9 = \beta(\beta y_7 + \varepsilon_8) + \varepsilon_9 &= \beta(\beta(\beta y_6 + \varepsilon_7) + \varepsilon_8) + \varepsilon_9 \\ &= \beta^3 y_6 + \beta^2 \varepsilon_7 + \beta\varepsilon_8 + \varepsilon_9. \end{aligned}$$

We have that ε_k does not depend on y_m, $m < k$. Hence $\mathbf{E}\{y_9|y_0, y_1, \dots, y_6\} = \beta^3 y_6 = 8y_6$.

Further, $\{y_k\}_{k \le 6}, \varepsilon_7, \varepsilon_8, \varepsilon_9$ are mutually independent. Hence $\beta^3 y_6, \beta^2\varepsilon_7, \beta\varepsilon_8, \varepsilon_9$ are mutually independent, and $\beta^3 y_6, \beta^2\varepsilon_7, \beta\varepsilon_8, \varepsilon_9$ are mutually independent given $y_1, \dots, y_6$. Hence

$$\begin{aligned} &\mathbf{E}\{y_9^2|y_0, y_1, \dots, y_6\} \\ &\quad = \beta^6 y_6^2 + \beta^4 \mathbf{E}\{\varepsilon_7^2|y_0, y_1, \dots, y_6\} + \beta^2 \mathbf{E}\{\varepsilon_8^2|y_0, y_1, \dots, y_6\} \\ &\qquad + \mathbf{E}\{\varepsilon_9^2|y_0, y_1, \dots, y_6\} \\ &\quad = \beta^6 y_6^2 + \beta^4 \mathbf{E}\varepsilon_7^2 + \beta^2 \mathbf{E}\varepsilon_8^2 + \mathbf{E}\varepsilon_9^2 \\ &\quad = 64y_6^2 + \sigma^2[16 + 4 + 1] = 64y_6^2 + \tfrac{1}{2}[16 + 4 + 1] = 64y_6^2 + 10.5. \qquad \square \end{aligned}$$

Estimation of the probability that the error is in a given interval

Assume that ε_{t+s} are normally distributed as i.i.d. $N(0, \sigma^2)$. Then y_{t+s} is normal with variance v_s conditionally given $y_0, y_1, \dots, y_t$. In other words, y_{t+s} is normal with the conditional variance

$$v_s^2 \triangleq \mathrm{Var}_t y_{t+s},$$

i.e., with the variance on the conditional probability space given $y_0, y_1, \dots, y_t$.

Recall that if $\xi \sim N(0,1)$, then $\mathbf{P}(|\xi| < x) = \Phi(x) - \Phi(-x)$. Further, if $\xi \sim N(0,\hat{v}^2)$ then $\xi/\hat{v} \sim N(0,1)$ and $\mathbf{P}(|\xi| < v \cdot x)\mathbf{P}(|\xi/\hat{v}| < x) = \Phi(x) - \Phi(-x)$. Hence we can estimate the conditional probability that the absolute value of the difference between y_{t+s} and its forecast $\mathbf{E}_t y_{t+1}$ is less than a given $\tilde{x} = v_s \cdot x$:

$$\mathbf{P}_t\left(|y_{t+s} - \mathbf{E}_t y_{t+s}| < v_s \cdot x\right) = \Phi(x) - \Phi(-x),$$

where $\mathbf{P}_t \triangleq \mathbf{P}(\cdot\,|\,y_0, y_1, \ldots, y_t)$, or

$$\mathbf{P}_t\left(|y_{t+s} - \mathbf{E}_t y_{t+s}| < x\right) = \Phi(x/v_s) - \Phi(-x/v_s).$$

Confidence interval

Let us consider the following problem: find an interval that contains y_{t+s} with a given probability p.

Again, we need to find x such that $\Phi(x) - \Phi(-x) = p$.

For instance, if $\xi \sim N(0,\hat{v}^2)$ then $\mathbf{P}(|\xi/\hat{v}| < 1.65) = \Phi(1.65) - \Phi(-1.65) = 0.9$, $\mathbf{P}(|\xi/\hat{v}| < 2) = \Phi(2) - \Phi(-2) = 0.95$.

Hence

$$\mathbf{P}_t\left(|y_{t+s} - \mathbf{E}_t y_{t+s}| < 1.65 \cdot \hat{v}_s\right) = 0.9, \quad \mathbf{P}_t\left(|y_{t+s} - \mathbf{E}_t y_{t+s}| < 2 \cdot v_s\right) = 0.95.$$

By this method, we can define confidence intervals around the point forecasts. For instance, the interval $\pm 1.65 v_s$ gives a 90% confidence interval.

Forecasting with unknown (β_0, β)

The most straightforward approach is to find an estimate $(\hat{\beta}_0, \hat{\beta})$ of (β_0, β) (for instance, by the LS estimate method), and then apply the forecasting rule with this $(\hat{\beta}_0, \hat{\beta})$. The same approach can be applied to estimate the confidence interval. Clearly, the replacement of (β_0, β) for $(\hat{\beta}_0, \hat{\beta})$ generates an additional error, but we are not going to discuss its impact here.

Problem 8.31 Forecast the fourth and fifth terms in the sequence 1, −1, 0.5, assuming the autoregression model

$$y_k = \beta y_{k-1} + \varepsilon_k, \qquad k = 1, 2, \ldots,$$

where ε_k is the white noise process, $\mathbf{E}\varepsilon_t^2 = \sigma^2$, and where β is unknown.

Solution. Our model is a special case of the regression model

$$y_k = \beta x_k + \varepsilon_k, \quad \text{where} \quad (x_1, x_2) = (1, -1), \quad (y_1, y_2) = (-1, 0.5).$$

By the formula from Chapter 2 (the case of $\beta_0 = 0$), the LS estimate of β is

$$\hat{\beta} = \frac{\sum_{i=1}^{T} x_i y_i}{\sum_{i=1}^{T} x_i^2} = \frac{-1 - 0.5}{1 + 1} = -\frac{3}{4}, \qquad T = 2.$$

We have proved above that $\mathbf{E}\{y_{k+s}|y_0, y_1, \dots, y_k\} = \beta^s y_k$. We do not know β, so we are in a position to replace it by its estimate $\hat{\beta}$. Then the forecast is

$$\mathbf{E}\{y_{k+s}|y_0, y_1, \dots, y_k\} \sim \hat{\beta}^s y_k,$$

i.e.,

$$\mathbf{E}\{y_3|y_0, y_1, y_2\} \sim \hat{\beta} y_2 = 0.5 \cdot (-3/4) = -3/8 = -0.375,$$

$$\mathbf{E}\{y_4|y_0, y_1, y_2\} \sim \hat{\beta}^2 y_2 = 0.5 \cdot (-3/4)^2 = 9/32 = 0.2813. \quad \square$$

Problem 8.32 Using historical daily prices for 100 days and a MATLAB program, forecast the return for N days. Estimate an interval such that it contains the increments of the log of prices (and/or prices) for 102, 110, 200 days with probability 0.9.

Solution is represented by the following program. (It is assumed that the prices are stored in the ASCII file ANZ.mat and are placed in the working directory; this file contains a matrix, where the second column represents the prices under consideration.)

MATLAB code for the forecast of price return

```
function[forecast,Confid_interval]=forecast06(m,N,p)
%m - number of days observed; m=100; N day forecasts
%p - probability; p=0.9

load anz -ascii; yyy=anz; yy=zeros(m,1); xx=yy; m1=m-2;
y=ones(m1,1); x=y;
 one=ones(m1,1); b=1; t=one;
for k=1:m
    yy(k)=log(yyy(k,2));
end; for k=1:m1 t(k)=k;
    y(k)=yy(k+2)-yy(k+1); x(k)=yy(k+1)-yy(k);
end;

b=(sum(y .*x)-sum(y .*one)*sum(x .*one)/m1)/(sum(x .*x)-sum(x
.*one)^2/m1); b0=sum(y .*one)/m1-b*sum(x .*one)/m1;

forecast=y(m1)*(b^N)+b0*(1-b^(N-1))/(1-b); kk=m1+N;
TrueValue=log(yyy(k+2,2))-log(yyy(k+1,2)) u=b0*ones(m1,1)+b*x-y;
```

```
sigmaLS=sqrt(sum(u .*u)/(m1-1)); x=interval(0.9,1000)
Confid_interval=x*sqrt((1-b^(2*N))/(1-b^2))*sigmaLS
```

Remark 8.33 Typically, the reliability of forecasts for financial data is relatively low, since the trend is usually overshadowed by the volatility. In addition, the stationarity conditions do not usually hold for financial data, and there is no guarantee that a given evolution law holds for long enough. In Problem 8.32, the forecast is a conditional expectation calculated for a certain model for autoregression. Similar forecast algorithms are used in practice (with more sophisticated models and with a larger number of parameters). These and other methods of forecast for financial data are intensively studied. A very popular approach for financial forecast is so-called *technical analysis* (which is usually considered as a non-academic approach); the algorithms in this framework are based on statistical methods as well as on empirical rules.

Forecasting with an AR(p)

All the previous calculations can be extended for a pth-order autoregressive process, AR(p)

$$y_t = \beta_1 y_{t-1} + \beta_2 y_{t-2} + \cdots + \beta_p y_{t-p} + \varepsilon_t.$$

As an example, consider a forecast of y_{t+1} based on the information in t by using an AR(2)

$$y_{t+1} = \beta_1 y_t + \beta_2 y_{t-1} + \varepsilon_{t+1}. \tag{8.39}$$

This immediately gives the one-period point forecast

$$\mathbf{E}_t y_{t+1} = \beta_1 y_t + \beta_2 y_{t-1}. \tag{8.40}$$

We can use (8.39) to write y_{t+2} as

$$\begin{aligned} y_{t+2} &= \beta_1 y_{t+1} + \beta_2 y_t + \varepsilon_{t+2} \\ &= \beta_1(\beta_1 y_t + \beta_2 y_{t-1} + \varepsilon_{t+1}) + (\beta_1^2 + \beta_2) y_t + \beta_1 \beta_2 y_{t-1} \\ &\quad + \beta_1 \varepsilon_{t+1} + \varepsilon_{t+2}. \end{aligned} \tag{8.41}$$

Similarly, formulae for forecasts and forecast error variances can be obtained for $p > 2$.

8.10 Heteroscedastic residuals, ARCH and GARCH

Suppose we have estimated a regression model, and obtained the fitted residuals u_t for $t = 1, \dots, T$, and we are not sure if the errors are homoscedastic, i.e., if the conditions of Definition 8.20 are satisfied for ε_t. The real errors ε_t are non-observable, so we may try to verify whether u_t are homoscedastic, or if $\mathbf{E}\{u_t^2 \mid \mathcal{F}_{t-1}\} = \text{const.}$ for all t, where $\mathcal{F}_t$ is the filtration generated by the observations.

Definition 8.34 *If the conditions of Definition 8.20 are not satisfied for u_t, i.e., $\mathbf{E}\{u_t^2 \mid \mathcal{F}_{t-1}\} \neq \text{const.}$ for all t, we say that the residuals u_t are heteroscedastic.*

Heteroscedasticity and homoscedasticity are important characteristics of statistical data, and the related analysis is crucial for many models. For instance, stock price models with time-varying random volatility lead to heteroscedastic models, and heteroscedasticity may have an impact on hypothesis testing in the LS setting described above. However, the LS method is *consistent* (converges to the true value as the sample size increases) even if the residuals are heteroscedastic.

8.10.1 *Some tests of heteroscedasticity*

In fact, heteroscedasticity can be detected by testing the autocorrelation of u_t^2; if there is autocorrelation, then the residuals are heteroscedastic.

Engles test

Consider the AR(q) model for residual

$$u_t^2 = \beta_0 + \sum_{s=1}^{q} \beta_i u_{t-s}^2 + v_t.$$

Then one can test the hypothesis that $\beta_i = 0$ for $i = 1, \dots, q$. (If it is not true, then the residuals are autocorrelated and heteroscedastic.)

Durbin–Watson test

The classic test for autocorrelation is the Durbin–Watson test. It requires us to find

$$\mu_{DW} = \frac{\sum_{t=2}^{T} (u_t - u_{t-1})^2}{\sum_{t=1}^{T} u_t^2}.$$

Clearly,

$$\mu_{DW} = \frac{\sum_{t=2}^{T} u_t^2 + \sum_{t=2}^{T} u_{t-1}^2 - 2\sum_{t=2}^{T} u_t u_{t-1}}{\sum_{t=1}^{T} u_t^2}.$$

For large T, the differences in the ranges of the summations in this expression can be ignored, and

$$\mu_{DW} \sim 2(1-\phi), \quad \phi \triangleq \frac{\sum_{t=1}^{T} u_t u_{t-1}}{\sum_{t=1}^{T} u_t^2}.$$

We have that $\phi \in [-1, 1]$. It follows that μ_{DW} ranges through $[0, 4]$ (for large T).

The following criteria are commonly accepted:

- If $\mu_{DW} < 2$ for positive autocorrelation;
- If $\mu_{DW} > 2$ for negative autocorrelation;
- If $\mu_{DW} \sim 2$ for zero autocorrelation.

Clearly, one may use the value ϕ instead of μ_{DW}; for instance, $\phi \sim 0$ means zero autocorrelation.

In fact, exact critical values cannot be computed and instead an upper bound and a lower bound are used to test the null or zero autocorrelation.

8.10.2 ARCH models

Let us describe the Autoregressive Conditional Heteroscedastic Model (ARCH) using as an example the so-called ARCH(1) model. Consider first the following regression:

$$y_t = x_t^\top \beta + \varepsilon_t, \quad t = 0, 1, 2, \dots. \tag{8.42}$$

Here y_t and x_t are observable values, $y_t \in \mathbf{R}$, $x_t \in \mathbf{R}^k$. The vector $\beta \in \mathbf{R}^k$ is a non-observable parameter, ε_t are non-observable errors.

We call this model the ARCH(1) model since σ_t^2 depends only on ε_{t-1}.

Let $\mathbf{E}_t$ denote the conditional expectation given observations $y_0, y_1, \dots, y_t$, $x_0, x_1, \dots, x_t$. Let Var_t denote the corresponding conditional variance.

Further, we assume that

$$\varepsilon_t = \eta_t \sigma_t, \dots, \quad \mathbf{E}_{t-1}\eta_t = 0, \quad \mathbf{E}_{t-1}\eta_t^2 = 1, \quad \text{with} \quad t = 1, 2, \dots, \tag{8.43}$$

where η_t is the normalized error, and σ_t^2 is the conditional variance $\sigma_t^2 = \mathrm{Var}_{t-1}\varepsilon_t$ (conditional given observations $y_0, y_1, \dots, y_t, x_0, x_1, \dots, x_t$).

Finally, we assume the following evolution law for σ_t^2:

$$\sigma_t^2 = \alpha_0 + \alpha_1 \varepsilon_{t-1}^2, \quad \alpha_0 > 0, \quad \alpha_1 \in [0, 1). \tag{8.44}$$

Here α_0 and α_1 are some parameters that are usually unknown and need to be estimated from statistical data.

By (8.43), it follows that

$$\mathbf{E}_{t-1}\varepsilon_t = 0 = \sigma_t \mathbf{E}_{t-1}\eta_t = 0, \quad \sigma_t^2 = \mathbf{E}_{t-1}\varepsilon_t^2. \tag{8.45}$$

It is clear that the unconditional distribution of ε_t is non-normal even if we assume that η_t are i.i.d. $N(0, 1)$.

Note that the conventions for the time subscript of the conditional variance can be different. Some authors use σ_{t-1}^2 to indicate the variance in t.

Starting value of σ^2

Note that we need a starting value of $\sigma_1^2 = \alpha_0 + \alpha_0\varepsilon_0^2$. To obtain σ_1, we need ε_0. It is why we have a sample running from $t = 0$ to $t = T$, but observation $t = 0$ is only used to construct a starting value of σ_1^2.

Restrictions on α_0 and α_1

To guarantee that $\sigma_t^2 > 0$, we require that

$$\alpha_0 \geq 0, \quad \alpha_1 \geq 0, \quad \alpha_0 + \alpha_1 > 0.$$

In addition, we require that

$$\alpha_1 < 1$$

to ensure that the unconditional variance is bounded as $t \to +\infty$. To see this, use (8.45) to write (8.44) as

$$\mathbf{E}_{t-1}\varepsilon_t^2 = \alpha_0 + \alpha_1\varepsilon_{t-1}^2, \tag{8.46}$$

which implies

$$\mathbf{E}\varepsilon_t^2 = \alpha_0 + \alpha_1\mathbf{E}\varepsilon_{t-1}^2.$$

Set $V_t \triangleq \mathbf{E}\varepsilon_t^2$. We have that this process evolves as a deterministic dynamic system

$$V_t = \alpha_0 + \alpha_1 V_{t-1}, \quad t = 0, 1, 2, \ldots.$$

Clearly, this equation has a bounded solution if and only if $|\alpha_1| < 1$. It can be seen from

$$V_t = \alpha_0 + \alpha_1(\alpha_0 + \alpha_1\mathbf{E}\varepsilon_{t-2}^2) = \cdots = \alpha_0(1 + \alpha_1 + \alpha_1^2 + \cdots + \alpha_1^t\mathbf{E}\varepsilon_0^2).$$

Forecasting of ε_t^2 ***and*** σ_t^2 ***for known*** (α_0, α_1)

By (8.45), $\sigma_{t+s}^2 = \mathbf{E}_{t+s-1}\varepsilon_{t+s}^2$, $s = 0, 1, 2, \ldots$ Hence

$$\mathbf{E}_{t-1}\sigma_{t+s}^2 = \mathbf{E}_{t-1}\mathbf{E}_{t+s-1}\varepsilon_{t+s}^2, \qquad s = 0, 1, 2, \ldots.$$

Hence

$$\mathbf{E}_{t-1}\sigma_{t+s}^2 = \mathbf{E}_{t-1}\varepsilon_{t+s}^2, \qquad s = 0, 1, 2, \ldots.$$

Therefore, it suffices to describe forecasting of ε_t^2. Similarly to (8.46), we have

$$\mathbf{E}_t\varepsilon_{t+1}^2 = \alpha_0 + \alpha_1\varepsilon_t^2.$$

By (8.46),

$$\mathbf{E}_{t-1}\varepsilon_{t+1}^2 = \alpha_0 + \alpha_1\mathbf{E}_{t-1}\varepsilon_t^2 = \alpha_0(1 + \alpha_1) + \alpha_1^2\varepsilon_{t-1}^2. \tag{8.47}$$

It follows that

$$\mathbf{E}_{t-1}\varepsilon_{t+2}^2 = \alpha_0 + \alpha_1\mathbf{E}_{t-1}\varepsilon_{t+1}^2 = \alpha_0(1 + \alpha_1 + \alpha_1^2) + \alpha_1^3\varepsilon_{t-1}^2. \tag{8.48}$$

Continuation of this process gives

$$\begin{aligned}\mathbf{E}_{t-1}\varepsilon_{t+s}^2 = \mathbf{E}_{t-1}\sigma_{t+s}^2 = \mathbf{E}_{t-1}\varepsilon_{t+s}^2 &= \alpha_0\left(\frac{1}{1-\alpha_1} - \frac{\alpha_1^s}{1-\alpha_1}\right) + \alpha_1^{s+1}\varepsilon_{t-1}^2 \\ &= \frac{\alpha_0}{1-\alpha_1} + \alpha_1^{s+1}\left(\varepsilon_{t-1}^2 - \frac{\alpha_0}{1-\alpha_1}\right).\end{aligned} \tag{8.49}$$

Remark 8.35 If $|\alpha_1| < 1$ then, for any ε_{t-1},

$$\mathbf{E}_{t-1}\varepsilon_{t+s}^2 \to \frac{\alpha_0}{1-\alpha_1}, \qquad s \to +\infty.$$

Problem 8.36 Prove Remark 8.35.

8.10.3 Estimation of parameters for ARCH(1) with the ML method

The most common way of estimating the model is to assume that η_t are i.i.d. with law $N(0, 1)$ and to set up the likelihood function. The log-likelihood is easily found, since the model is conditionally Gaussian.

Remember that $\sigma_t^2 = \alpha_0 + \alpha_1\varepsilon_{t-1}^2$. This means that if ε_{t-1}^2 is given, then σ_t^2 is given (for known non-random α_0 and α_1). On the other hand, the value of σ_{t-1}^2 is the only information from $(\varepsilon_{t-1}, \varepsilon_{t-2}, \ldots, \varepsilon_0)$ that can have an impact on ε_t. Therefore, the conditional probability density function of ε_t given $(\varepsilon_{t-1}, \varepsilon_{t-2}, \ldots, \varepsilon_0)$ is

$$p_{\varepsilon_t}(e|\varepsilon_{t-1}, \varepsilon_{t-2}, \ldots, \varepsilon_0) = p_{\varepsilon_t}(e|\varepsilon_{t-1}) = p_{\varepsilon_t}(e|\sigma_t^2) = \frac{1}{\sqrt{2\pi\sigma_t^2}}\exp\left(-\frac{e^2}{2\sigma_t^2}\right).$$

Thus, we get the joint probability density function of the $(\varepsilon_1, \varepsilon_2, \ldots, \varepsilon_T)$ (conditional given ε_0) by multiplying the conditional probability density functions of each of the errors

$$
\begin{aligned}
p_{\varepsilon_1,\ldots,\varepsilon_T}(e_1,\ldots,e_T\,|\,\varepsilon_0) &= p_{\varepsilon_1}(e_1|\varepsilon_0)p_{\varepsilon_2}(e_2|\varepsilon_1,\varepsilon_0)\times\cdots\times p_{\varepsilon_T}(e_T|\varepsilon_{T-1},\ldots,\varepsilon_0)\\
&= p_{\varepsilon_1}(e_1|\varepsilon_0)p_{\varepsilon_2}(e_2|\sigma_2^2)\times\cdots\times p_{\varepsilon_T}(e_T|\sigma_T^2)\\
&= \prod_{t=1}^{T}\frac{1}{\sqrt{2\pi\sigma_t^2}}\exp\left(-\sum_{t=1}^{T}\frac{e_t^2}{2\sigma_t^2}\right).
\end{aligned}
$$

It is more convenient to use the logarithm of the probability density function, or the log-likelihood function

$$
L(e_1,\ldots,e_T) = \ln p_{\varepsilon_1,\ldots,\varepsilon_T}(e_1,\ldots,e_T) = -\frac{T}{2}\ln(2\pi) - \frac{1}{2}\sum_{t=1}^{T}\ln\sigma_t^2 - \sum_{t=1}^{T}\frac{e_t^2}{2\sigma_t^2}.
$$

Assume that we are given the observations $(x_0, x_1, \ldots, x_T, y_0, y_1, \ldots, y_T)$. We are looking for a set of estimates $(\hat{\beta}, \hat{\alpha}_0, \hat{\alpha}_1)$ of unknown parameters $(\beta, \alpha_0, \alpha_1)$.

The ML estimation method requires us to find the value of $(\hat{\beta}, \hat{\alpha}_0, \hat{\alpha}_1)$ that maximizes $L(u_1, \ldots, u_T)$ given the observations $(x_0, x_1, \ldots, x_T, y_0, y_1, \ldots, y_T)$.

Therefore, we need to maximize over $(\hat{b}, \hat{\alpha}_0, \hat{\alpha}_1)$

$$
L(u_1, \ldots, u_T) = -\frac{T}{2}\ln(2\pi) - \frac{1}{2}\sum_{t=1}^{T}\ln\hat{\sigma}_t^2 - \sum_{t=1}^{T}\frac{u_t^2}{2\hat{\sigma}_t^2}, \tag{8.50}
$$

where the fitted residuals and their fitted variances are defined by $(\hat{b}, \hat{\alpha}_0, \alpha_0)$ and by the observations as

$$
u_t \triangleq y_t - x_t^\top\hat{\beta}, \quad t = 0, 1, 2, \ldots \quad \hat{\sigma}_t^2 \triangleq \hat{\alpha}_0 + \hat{\alpha}_1 u_{t-1}^2, \quad t = 1, 2, \ldots. \tag{8.51}
$$

The likelihood function L is maximized by minimizing the last two terms in (8.50). Substitute values from (8.51) for u_t and σ_t^2, then L is a function of $(\hat{\beta}, \alpha_0, \alpha_1)$. Then L is maximized with respect to the parameters. This requires some optimization methods.

Note that ML methods give the same results as an LS with weight coefficient described in Remark 8.10 (with errors replaced by residuals).

Problem 8.37 Consider the following model:

$$
\begin{aligned}
&y_t = \beta y_{t-1} + \varepsilon_t, \quad t = 0, 1, 2, \ldots, \qquad \mathbf{E}\{\varepsilon_t|x_t\} = 0,\\
&\varepsilon_t = \eta_t\sigma_t \quad \text{with} \quad t = 0, 1, 2, \ldots, \qquad \mathbf{E}_{t-1}\eta_t = 0, \quad \mathbf{E}_{t-1}\eta_t^2 = 1,\\
&\sigma_t^2 = \alpha_0 + \alpha_1\varepsilon_{t-1}^2, \qquad \alpha_0 \geq 0, \quad \alpha_1 \in [0, 1), \quad \alpha_0 + \alpha_1 > 0.
\end{aligned}
\tag{8.52}
$$

Here y_t are observable values, $y_t \in \mathbf{R}$, and the parameter $(\beta, \alpha_0, \alpha_1) \in \mathbf{R}^3$ is non-observable. $\mathbf{E}_t$ denotes the conditional expectation given observations $\{y_s\}_{s \le t}$. In addition, we assume that η_t are i.i.d. $N(0,1)$, and η_t does not depend on $\{y_s\}_{s \le t-1}$.

Assume that we are given a series of data $y_k = \ln S(t_k) - \ln S(t_{k-1})$ for quarterly prices $(S(t_0), S(t_1), S(t_2), S(t_3)) = (e^0, e^{1.1}, e^{3.1}, e^{4.6})$, where $(t_0, t_1, t_2, t_3) =$ (January 1st, April 1st, July 1st, October 1st). Find out which set of parameters is more likely to be the true one: $(\hat{\beta}, \hat{\alpha}_0, \hat{\alpha}_1) = (0.1, 0.25, 0)$ or $(\hat{\beta}, \hat{\alpha}_0, \hat{\alpha}_1) = (0.1, 0.05, 0.5)$.

Solution. We are given a series of data $\{y_t\} = \{1.1, 2, 1.5\}$. Apply the algorithm described above for the case when $x_t = y_{t-1}$. We have that:

- u_1 is defined by (x_1, y_1), or by (y_0, y_1);
- $\hat{\sigma}_1$ is defined by u_0;
- u_0 is defined by (x_0, y_0), or by (y_{-1}, y_0).

Therefore, we need terms y_{-1} and y_0 to proceed. Since we have only three terms, we must take $T = 1$ and $(y_{-1}, y_0, y_1) = (1.1, 2, 1.5)$.

We have

$$u_1 = 1.5 - 2\hat{\beta}, \quad u_0 = 2 - 1.1\hat{\beta}, \quad \hat{\sigma}_1^2 = \hat{\alpha}_0 + \hat{\alpha}_1 u_0^2.$$

If $(\hat{\beta}, \hat{\alpha}_0, \hat{\alpha}) = (0.1, 0.25, 0)$, then

$$u_1 = 1.5 - 2\hat{\beta} = 1.3, \quad u_0 = 2 - 1.1\hat{\beta} = 1.89,$$

$$\hat{\sigma}_1^2 = \hat{\alpha}_0 + \hat{\alpha} u_0^2 = 0.25 + 0 \cdot 1.89^2 = 0.25,$$

and

$$L(u_1, \dots, u_T) + \frac{T}{2}\ln(2\pi) = L(u_1) + \frac{1}{2}\ln(2\pi) = -\frac{1}{2}\sum_{t=1}^{T}\ln\hat{\sigma}_t^2 - \sum_{t=1}^{T}\frac{u_t^2}{2\hat{\sigma}_t^2}$$

$$= -\frac{1}{2}\ln\hat{\sigma}_1^2 - \frac{u_1^2}{2\hat{\sigma}_1^2} = -\frac{1}{2}\ln 0.25 - \frac{1.3^2}{2 \cdot 0.25}$$

$$= -2.687.$$

Similarly, if $(\hat{\beta}, \hat{\alpha}_0, \hat{\alpha}_1) = (0.1, 0.05, 0.5)$, then

$$u_1 = 1.5 - 2\hat{\beta} = 1.3, \quad u_0 = 2 - 1.1\hat{\beta} = 1.89,$$

$$\hat{\sigma}_1^2 = \hat{\alpha}_0 + \hat{\alpha}_1 u_0^2 = 0.05 + 0.5 \cdot 1.89^2 = 1.836,$$

and

$$
\begin{aligned}
L(u_1, \ldots, u_T) + \frac{T}{2}\ln(2\pi) &= L(u_1) + \frac{1}{2}\ln(2\pi) \\
&= -\frac{1}{2}\ln\hat{\sigma}_1^2 - \frac{u_1^2}{2\hat{\sigma}_1^2} \\
&= -\frac{1}{2}\ln 1.836 - \frac{1.3^2}{2 \cdot 1.8361} = -0.7640.
\end{aligned}
$$

Thus, the set of parameters $(\hat{\beta}, \hat{\alpha}_0, \hat{\alpha}_1) = (0.1, 0.05, 0.5)$ is more likely to be the true set.

In addition, if one finds that all other hypotheses with $\hat{\alpha}_1 = 0$ are less likely than the ones with $\hat{\alpha}_1 \neq 0$ then it may be concluded that the right model is heteroscedastic (of course, the reliability of this conclusion is low for this particular Problem 8.37, since the size of the statistical series is small).

The solution of Problem 8.37 shows how to use optimization techniques for maximization of the likelihood function, since it is clear now how to form the optimality criteria.

8.10.4 ARCH(q) and GARCH models

The ARCH(q) is the regression (8.42) together with

$$
\sigma_t^2 = \alpha_0 + \alpha_1\varepsilon_{t-1}^2 + \alpha_1\varepsilon_{t-2}^2 + \cdots + \alpha_q\varepsilon_{t-q}^2, \quad \alpha_0 > 0, \quad \alpha_k \in [0, 1). \tag{8.53}
$$

The GARCH(1,1) (generalized ARCH) model is the regression (8.42) together with

$$
\begin{aligned}
&\sigma_t^2 = \alpha_0 + \alpha_1\varepsilon_{t-1}^2 + \beta_1\sigma_{t-2}^2, \\
&\alpha_0 > 0, \quad \alpha_1 \geq 0, \quad \beta_1 \geq 0, \quad \alpha_1 + \beta_1 < 1.
\end{aligned} \tag{8.54}
$$

Restrictions for parameters α_0, α_2, β_1 for GARCH(1,1)

To ensure that $\sigma_t^2 \geq 0$, we need to take $\alpha_0 \geq 0$, $\alpha_2 \geq 0$, $\beta_1 \geq 0$.

Let us show that the restriction $\alpha_1 + \beta_1 < 1$ ensures that process σ_t^2 is such that the unconditional variance is finite if $\mathbf{E}\varepsilon_t^2$ is bounded as $T \to \infty$. To see this, note that

$$
\sigma_t^2 = \mathbf{E}_{t-1}\varepsilon_t^2.
$$

Hence

$$
\mathbf{E}_{t-1}\varepsilon_t^2 = \alpha_0 + \alpha_1\varepsilon_{t-1}^2 + \beta_1\mathbf{E}_{t-2}\varepsilon_{t-1}^2.
$$

Set $V_t \triangleq \mathbf{E}\varepsilon_t^2$. We have that

$$V_t = \alpha_0 + (\alpha_1 + \beta_1)V_{t-1}, \quad t \to +\infty.$$

Clearly, this equation has a bounded solution if and only if $|\alpha_1 + \beta_1| < 1$. In this case

$$V_t = \mathbf{E}\varepsilon_t^2 \to \frac{\alpha_0}{1 - \alpha_1 - \beta_1} \quad \text{as} \quad t \to +\infty. \tag{8.55}$$

The GARCH(1,1) corresponds to an ARCH(∞)

Solution of (8.54) recursively by substituting for σ_{t-s}^2 gives

$$\begin{aligned}\sigma_t^2 &= \alpha_0 + \alpha_1\varepsilon_{t-1}^2 + \beta_1(\alpha_0 + \alpha_1\varepsilon_{t-2}^2 + \beta_1\sigma_{t-2}^2)\\ &= \alpha_0(1 + \beta_1) + \alpha_1\varepsilon_{t-1}^2 + \beta_1\alpha_1\varepsilon_{t-2}^2 + \beta_1^2\sigma_{t-2}^2\\ &= \cdots = \frac{\alpha_0}{1 - \beta_1} + \alpha_1\sum_{j=0}^{+\infty}\beta_1^j\varepsilon_{t-1-j}^2.\end{aligned}$$

Therefore, the GARCH(1,1) model can be interpreted as an ARCH model with unlimited memory (with exponentially declining coefficients for long memory, since $0 \le \beta_1 < 1$).

Forecasting with GARCH(1,1)

Similarly to the case of the ARCH model (8.49), but with the sum of α_1 and β_1 as the AR(1) parameter, repeated substitutions show that

$$\mathbf{E}_{t-1}\varepsilon_{t+s}^2 = \frac{\alpha_0}{1 - \alpha_1 - \beta_1} + (\alpha_1 + \beta_1)^{s+1}\left(\varepsilon_{t-1}^2 - \frac{\alpha_0}{1 - \alpha_1 - \beta_1}\right). \tag{8.56}$$

It is the forecasting formula for errors.

Estimation of parameters

To estimate the model consisting of (8.42) and (8.54) we can still use the likelihood function as was described for the ARCH model (with some adjustment). We need to create the starting value of ε_0 as in the ARCH model (use y_0 and x_0 to create ε_0), but for GARCH(1,1) we also need a starting value of σ_0^2. It is often recommended that we use $\sigma_0^2 = \text{Var}(u_t)$, where u_t are the residuals from an LS estimation of (8.42) (clearly, this is applicable if the process is stationary).

GARCH(p,q)

The GARCH(p,q) is the regression (8.42) together with

$$\sigma_t^2 = \alpha_0 + \sum_{i=1}^{p} \beta_i \sigma_{t-i}^2 + \sum_{j=1}^{q} \alpha_j \varepsilon_{t-j}^2 + \beta_1,$$

where $\alpha_0 > 0,\ \alpha_j \geq 0,\ \beta_j \geq 0,\ \sum_{i=1}^{p} \beta_i + \sum_{j=1}^{q} \alpha_j < 1$.

8.11 Problems

Problem 8.38 Estimate the parameters (a, σ) by an LS estimator in the AR(1) model $y_t = \beta y_{t-1} + \varepsilon_t$, $\mathrm{Var}\, \varepsilon_t = \sigma^2$, for the series: (i) $(1, -2, 3, 1, 0, 0, 1)$; (ii) $(1, -1, 1, -1, 1, -1, 1)$.

Problem 8.39 Forecast the series with AR(1) model for k steps: (i) $(1, -2, 3, 1, 0, 0, 1)$; (ii) $(1, -1, 1, -1, 1, -1, 1)$. (*Hint:* it is the series from Exercise 8.38.)

Problem 8.40 Let $y_k = R(t_k)$, where $R(t) = \ln S(t_k)$, and where $S(t_k)$ are the daily prices for some stock, $k = 1, \ldots, N$, for some large enough N. (Find some prices on the internet or in newspapers.)

(i) Using the LS estimator, forecast y_k for $N + 2$, $N + 10$, $N + 100$ days. Use this result to forecast the corresponding $R(t)$.
(ii) Estimate intervals that contain these increments of log of prices (and/or prices) with probability 0.7.

9 Estimation of models for stock prices

In this chapter, methods of statistical analysis are applied to historical stock prices. We show how to estimate the appreciation rate and the volatility for some continuous time stock price models. Some generic methods of forecast of evolution of prices and parameters are also given.

9.1 Review of the continuous time model

Let us consider again the stock price equation

$$dS(t) = a(t)S(t)dt + \sigma(t)S(t)dw(t), \tag{9.1}$$

where $a(t)$ is the appreciation rate, $\sigma(t)$ is the volatility, $w(t)$ is a Wiener process, $w(t) \sim N(0, t)$.

Remember that

$$S(T) = S(0)e^{R(t)}, \quad R(t) \triangleq \int_0^t \mu(s)ds + \int_0^t \sigma(s)dw(s). \tag{9.2}$$

Let us repeat the proof that was given above in a different form. Let $f(x) = e^x$. By the Ito formula, it follows that

$$\begin{aligned} d_t f(R(t)) &= \frac{df}{dx}(R(t))dR(t) + \frac{1}{2}\frac{d^2f}{dx^2}(R(t))\sigma^2 dt \\ &= f(R(t))(\mu(t)dt + \sigma(t)dw(t) + \tfrac{1}{2}\sigma(t)^2 dt) \\ &= e^{R(t)}(\mu(t)dt + \sigma(t)dw(t) + \tfrac{1}{2}\sigma(t)^2 dt) \\ &= e^{R(t)}(a(t)dt + \sigma(t)dw(t)) = f(R(t))(a(t)dt + \sigma(t)dw(t)). \end{aligned}$$

Hence

$$d_t f(R(t)) = f(R(t))(a(t)dt + \sigma(t)dw(t)).$$

By uniqueness of the solution of the Ito equation for $S(t)$, it follows that $f(R(t)) = S(t)$ if $S(0) = f(R(0))$. Hence $f(R(t))/f(R(0)) = S(t)/S(0)$, and $R(t) = \ln[f(R(0))S(t)/S(0)] = \ln(S(t)/S(0)) + R(0)$. □

Usually, it is more convenient to apply statistical methods to the process $R(t)$, since it is more likely that its evolution has stationary characteristics.

Clearly, $R(t) \triangleq \ln(S(t)/S(0))$, $R(0) = 0$ and

$$dR(t) = \mu(t)dt + \sigma(t)dw(t), \tag{9.3}$$

where

$$\mu(t) \triangleq a(t) - \frac{\sigma(t)^2}{2}.$$

We can only observe prices as time series at a limited number of times, so we need to understand how the time series of prices is connected to the continuous time process (9.3).

9.2 Examples of special models for stock price evolution

The following two models are based on a certain hypothesis about (μ, σ).

The log-normal model without mean-reverting

For this model, μ and σ are non-random constants. Then the distribution of log price $R(T)$ at time T is normal with some mean μ and with variance $T\sigma^2$. In this case,

$$R(t) = R(t-h) + \mu h + q(t), \quad h > 0, \tag{9.4}$$

where

$$q(t) \triangleq \int_{t-h}^{t} \sigma dw(s) = \sigma[w(t) - w(t-h)].$$

Clearly,

$$\mathbf{E}q(t) = 0, \quad \mathbf{E}q(t)^2 = \mathbf{E}\left(\int_{t-h}^{t} \sigma dw(s)\right)^2 = \int_{t-h}^{t} \sigma^2 dt = \sigma^2 h,$$

and $q(t)$ is a Gaussian random variable distributed as $N(0, \sigma^2 h)$.

Moreover, if $t_k = hk, k = 0, 1, \ldots$, where $h > 0$, then $\{q(t_k)\}$ are mutually independent, and

$$R(t_k) = \beta R(t_{k-1}) + \beta_0 + \varepsilon_k,$$

where $\beta = 1$ and $\beta_0 = \mu h$, and $\{\varepsilon_k\}$ is a random discrete time process that represents stochastic changes. It is a Gaussian discrete time white noise process, $\varepsilon_k = q(t_k)$, $\mathbf{E}\varepsilon_k = 0$, and $\mathbf{E}\varepsilon_k^2 = \sigma^2 h$.

Problem 9.1 Show that if μ and σ are constants then

$$\begin{aligned} &R(t_2) - R(t_1) \sim N[\mu(t_2 - t_1), \sigma\sqrt{t_2 - t_1}], \\ &\mathbf{E}(R(t_2) - R(t_1))(R(t_4) - R(t_3)) = 0, \quad t_4 > t_3 > t_2 > t_1. \end{aligned} \tag{9.5}$$

Mean-reverting model

Let $\sigma > 0$ be a non-random constant. We say that $R(t)$ is a process with mean reverting if

$$dR(t) = (\alpha - \lambda R(t))dt + \sigma dw(t), \tag{9.6}$$

where $\sigma \neq 0$, $\alpha \in \mathbf{R}$, $\lambda > 0$ are some non-random constants. In this case, $\mu = \mu(t) \equiv \alpha - \lambda R(t)$.

The solution of the closed equation for $R(t)$ can be expressed as

$$R(t) = e^{-\lambda(t-h)}R(t-h) + \int_{t-h}^{t} e^{-\lambda(t-s)}\alpha ds + \int_{t-h}^{t} e^{-\lambda(t-s)}\sigma dw(s), \quad h > 0. \tag{9.7}$$

Hence

$$R(t) = e^{-\lambda h}R(t-h) + \lambda^{-1}(1 - e^{-\lambda h})\alpha + q(t), \quad h > 0. \tag{9.8}$$

Here $q(t) \triangleq \int_{t-h}^{t} e^{-\lambda(t-s)}\sigma dw(s)$ are Gaussian random variables, $\mathbf{E}q(t) = 0$,

$$\mathbf{E}q(t)^2 = \mathbf{E}\left(\int_{t-h}^{t} e^{-\lambda(t-s)}\sigma dw(s)\right)^2 = \int_{t-h}^{t} e^{-2\lambda(t-s)}\sigma^2 ds = \frac{1 - e^{-2\lambda h}}{2\lambda}\sigma^2.$$

Theorem 9.2 *For the mean-reverting model,* Var $R(t)$ *is bounded in* $t > 0$*, and* $R(t)$ *converges to a stationary process. More precisely, there exists a stationary* *process*[1] *$R_0(t)$ such that $\mathbf{E}|R(t) - R_0(t)|^2 \to 0$ as $t \to +\infty$. In fact, this process $R_0(t)$ can be expressed as*

$$R_0(t) = \int_{-\infty}^{t} e^{-\lambda(t-s)}\alpha ds + \int_{-\infty}^{t} e^{-\lambda(t-s)}\sigma dw(s) = \frac{\alpha}{\lambda} + \int_{-\infty}^{t} e^{-\lambda(t-s)}\sigma dw(s).$$

Note that we need a small modification of the definition of $w(t)$ and of the stochastic integral, to define $\int_{-\infty}^{0} \cdot dw(s)$. Normally, a Wiener process is defined on the time interval $[s_0, +\infty)$, where $s_0 \in \mathbf{R}$ is initial time; we had introduced Wiener processes for time interval $[0, +\infty)$ only. We can use the following definition: $\int_{-\infty}^{0} e^{-\lambda(t-s)}\sigma dw(s) \triangleq \int_{0}^{\infty} e^{-\lambda(t+s)}\sigma d\tilde{w}(s)$, where $\tilde{w}(s)$ is some standard Wiener process independent from $w(\cdot)$.

Proof of Theorem 9.2. To proof convergency, one should notice that R_0 satisfies

$$dR_0(t) = (\alpha - \lambda R_0(t))dt + \sigma dw(t), \quad R_0(0) = \frac{\alpha}{\lambda} + \int_{-\infty}^{0} e^{\lambda s}\sigma dw(s), \tag{9.9}$$

and $Y(t) \triangleq R_0(t) - R(t)$ satisfies

$$dY(t) = -\lambda Y(t)dt, \quad Y(0) = R_0(0), \tag{9.10}$$

i.e., $Y(t) = R_0(t) - R(t) = e^{-\lambda t}R_0(0)$. This process converges to zero in the required sense. □

If $t = t_1, t_2, \ldots$ and $h = t_k - t_{k-1}$, then $\{q(t_k)\}$ are mutually independent, and

$$R(t_k) = \beta R(t_{k-1}) + \beta_0 + \varepsilon_k,$$

where $\beta \triangleq e^{-\lambda h}$ and $\beta_0 \triangleq \lambda^{-1}(1 - e^{-\lambda h})\alpha$, and $\varepsilon_k \triangleq q(t_k)$ is a random discrete time process that represents stochastic changes. It is a discrete time white noise process, $\mathbf{E}\varepsilon_k = 0$, and $\mathbf{E}\varepsilon_k^2 = [(1 - e^{-2\lambda h})/(2\lambda)]\sigma^2$.

In the following chapters, we shall study estimation of market parameters from the observation of $\{R(t_k)\}$.

Note that the process $R(t)$ can be thought of as the limit of the discrete time process $R(t_k)$ as the time interval h becomes very small.

Remark 9.3 The stock price in the mean-reverting model has log-normal distribution as well as for the log-normal model. The log-normal model can be considered as a special case of the mean-reverting model with $\lambda = 0$.

1 A process $R_0(t)$ is said to be stationary if the distribution of the vector $(R(t_1 + h), \ldots, R(t_N + h))$ does not depend on the time shift h for any $N > 0$ and any set $(t_1, \ldots, t_N)$.

Another version of the mean-reverting model

For simplicity, assume that $\alpha = 0$. Then the model described above gives $S(t) = S(0)e^{R(t)}$ with $R(t) = \int_0^t e^{-\lambda(t-s)}\sigma dw(s)$, and this $R(t)$ converges to a stationary process with zero mean. In fact, this process oscillates near zero. On the other hand, the risk-free investment with initial wealth equal to $S(0)$ gives the total wealth equal to $\exp(\int_0^t r(s)ds)$. It is unnatural to expect that the risk-free investments have bigger systematic growth than risky stocks (in this case, it would be meaningless to invest in stocks). It would be more realistic to consider a model where the log prices oscillate near $\int_0^t r(s)ds$. This feature can be taken into account by the following model:

$$R(t) = \int_0^t r(s)ds + \tilde{R}(t),$$

$$d\tilde{R}(t) = (\alpha - \lambda\tilde{R}(t))dt + \sigma dw(t), \quad \tilde{R}(0) = 0,$$

where $\sigma \neq 0$, $\alpha \in \mathbf{R}$, and $\lambda > 0$ are some non-random constants, and where $r(t)$ is the currently observable risk-free interest rate. In this case, $\mu(t) \equiv r(t)+\alpha-\lambda\tilde{R}(t)$. Since $r(t)$ is an observable parameter and the process $R(t)$ is observable, we have that the process $\tilde{R}(t)$ is observable as well, and one may apply the method described above for the estimation of parameters $(\alpha, \lambda, \sigma)$ of the equation for $\tilde{R}(t)$.

Other models for stock prices

There are many other special types of diffusion models for stock prices:

- The volatility is non-random, and the appreciation rate $a(t)$ is an Ito process that evolves as

  $$da(t) = (\alpha - \lambda a(t))dt + \hat{\sigma} d\hat{w}(t),$$

  where $\hat{\sigma} \neq 0$, $\alpha \in \mathbf{R}$, $\lambda > 0$, and where $\hat{w}$ is some Wiener process;
- $(a(t), \sigma(t)) = f(\xi(t))$, where f is a deterministic function, ξ is a Markov chain process;
- $\sigma(t) = CS(t)^p$, where $C, p \in \mathbf{R}$;
- the volatility $\sigma(t)$ is an Ito process that evolves as

  $$d\sigma(t) = (\bar{\sigma} - \lambda\sigma(t))dt + \hat{\sigma} d\hat{w}(t),$$

  where $\hat{\sigma} \neq 0$, $\bar{\sigma} \in \mathbf{R}$, $\alpha \in \mathbf{R}$, $\lambda > 0$, and where $\hat{w}$ is some Wiener process.

All these models (and many others) need statistical evaluation in implementation with real market data, and that is one of the mainstream research fields in financial econometrics and statistical finance.

For a practical estimation of parameters of price models from historical data, we need statistical methods.

9.3 Estimation of models with constant volatility

Consider asset (stock) with the price $S(t)$ defined by (9.1). We shall use notations from Section 7.1. Up to the end of this section, we assume that the volatility σ is non-random and constant but its value is unknown a priori.

Remember that the Black–Scholes formula for the option price includes volatility and does not depend on μ (or on a) (Corollary 5.50).

9.3.1 Estimation of the log-normal model without mean-reverting

Consider the case when a, μ, and σ are constant. Then the distribution of log asset price $R(t)$ in T is normal with the variance $T\sigma^2$ and the mean μT (not important for the option price).

Let $t = t_1, t_2, \dots$ and $h = t_k - t_{k-1}$. We have found in Section 9.2 above that if $t = t_1, t_2, \dots$ and $h = t_k - t_{k-1}$, then

$$R(t_k) = \beta R(t_{k-1}) + \beta_0 + \varepsilon_k,$$

where $\beta = 1$ and $\beta_0 = \mu h$, and $\{\varepsilon_k\}$ is a random discrete time process that represents stochastic changes. It is a discrete time white noise process, $\varepsilon_k = q(t_k)$, $\mathbf{E}\varepsilon_k = 0$, and $\mathbf{E}\varepsilon_k^2 = \int_{t-h}^{t} \sigma^2 ds = \sigma^2 h$.

The LS estimator

By Proposition 8.12 and Definition 8.19, it can be seen that the LS estimator gives

$$\hat{\mu} h = \frac{1}{n} \sum_{k=1}^{n} (R(t_k) - R(t_{k-1})),$$

$$\hat{\sigma}^2 h = \frac{1}{n-1} \sum_{k=1}^{n} u_k^2, \quad \text{where } u_k \triangleq R(t_k) - R(t_{k-1}) - \hat{\mu} h, \tag{9.11}$$

and where $h = t_k - t_{k-1}$.

Example 9.4 If we have daily data, then $h = 1/365$ (or possibly something like $1/250$ if only weekdays are counted). If $\sigma^2 h$ estimated on daily data is $0.162/250$, then the volatility in the Black–Scholes formula is $\sigma^2 = 0.16$.

Of course, the estimate can change as new data points are added to the sample.

The ML estimator

By the results from Section 8.6, it can be seen that the ML estimator gives

$$\hat{\mu}h = \frac{1}{n}\sum_{k=1}^{n}(R(t_k) - R(t_{k-1})),$$

$$\hat{\sigma}^2 h = \frac{1}{n}\sum_{k=1}^{n}u_k^2, \quad \text{where } u_k = R(t_k) - R(t_{k-1}) - \hat{\mu}h, \tag{9.12}$$

and where $h = t_k - t_{k-1}$.

Note that the data for the two estimators described above can be obtained from the prices $S(t)$ directly as

$$u_k = \ln\frac{S(t_k)}{S(t_{k-1})} - \hat{\mu}h, \quad \hat{\mu}h = \frac{1}{n}\sum_{k=1}^{n}\ln\frac{S(t_k)}{S(t_{k-1})}.$$

Note that we do not assume that the sample mean is zero, even if this assumption is often made (to simplify the calculations).

Problem 9.5 Let the quarterly stock prices $(S(t_1),\ S(t_2),\ S(t_3),\ S(t_4)) = (e^0, e^{0.3}, e^{0.5}, e^{0.4})$ be given. Estimate parameters for the log-normal continuous time model without mean-reverting.

Solution. Note that $h = 1/4$. Take $y_k = R(t_k) = \ln S(t_k), x_k = y_{k-1}, k = 1, 2, 3,$ and apply the LS estimator with $T = 3$, when $\beta = 1$ is known. Then $\hat{\beta}_0 = \hat{\mu}h = (\sum_{k=1}^{3} y_k - \sum_{k=1}^{3} x_k)/3 = 0.1333$, hence $\hat{\mu} = 0.5333$. Further, we have fitted residuals

$$u_k = y_k - x_k - \hat{\mu}h, \qquad (u_1, u_2, u_3) = (0.1667, 0.0667, -0.2333),$$

hence the LS estimate $\hat{\sigma}$ for the volatility σ can be found from

$$\hat{\sigma}^2 h = \frac{1}{T-1}\sum_{k=1}^{T}u_k^2; \quad \hat{\sigma}^2 = 0.1733.$$

The estimate for the appreciation rate a is $\hat{a} = 0.62$. □

Let us show how the estimation of parameters described above can be applied for forecasting.

Problem 9.6 Under the assumptions and notations of the previous problem, find a forecast for the return $R(t_4)$ and $R(t_5)$ based on observations for times (t_0, t_1, t_3) and presuming that $t_k - t_{k+1}$ is three months for all k.

Solution. By (8.36), the desired forecast of $R(t_4)$ is $\mathbf{E}\{R(t_4)|y_0,y_1,y_2,y_3\} = R(t_3) + \mu h$. As was suggested in Section 8.9, we can replace this expectation by $R(t_3) + \hat{\mu}h = 0.4 + 0.1333 = 0.5333$. Similarly, (8.37) implies that the forecast of $R(t_5)$ is $R(t_3) + 2\hat{\mu}h = 0.4 + 0.2666 = 0.6666$. □

9.3.2 *Estimation of the mean-reverting model*

Let us estimate the parameters for the model

$$S(t) = S(0)e^{R(t)},$$

$$dR(t) = (\alpha - \lambda R(t))dt + \sigma dw(t), \quad R(0) = 0.$$

We have found in Section 9.2 that if $t = t_1, t_2, \dots$ and $h = t_k - t_{k-1}$, then

$$R(t_k) = \beta R(t_{k-1}) + \beta_0 + \varepsilon_k,$$

where $\beta \triangleq e^{-\lambda h}$ and $\beta_0 \triangleq \lambda^{-1}(1 - e^{-\lambda h})\alpha$, and $\{\varepsilon_k\}$ is a random discrete time process that represents stochastic changes. It is a discrete time white noise process, $\mathbf{E}\varepsilon_k = 0$, and $\mathbf{E}\varepsilon_k^2 = [(1 - e^{-2\lambda h})/2\lambda]\sigma^2$ (see (9.8)). Then estimates $\hat{\beta}$ and $\hat{\beta}_0$ can be found by LS or ML methods, as well as the LS or ML estimate of the variance (square) σ_{LS}^2 of $\mathbf{E}\varepsilon_k^2$. Then an estimate $(\hat{\alpha}, \hat{\lambda}, \hat{\sigma})$ of unknown parameters $(\hat{\alpha}, \hat{\lambda}, \hat{\sigma})$ can be found from the system

$$\begin{cases} e^{-\hat{\lambda}h} & = \hat{\beta} \\ \hat{\lambda}^{-1}(1 - e^{-\hat{\lambda}h})\hat{\alpha} & = \hat{\beta}_0 \\ \dfrac{1 - e^{-2\hat{\lambda}h}}{2\hat{\lambda}}\sigma^2 & = \sigma_{LS}^2 \end{cases} \quad \text{i.e.,} \quad \begin{cases} e^{-\hat{\lambda}h} & = \hat{\beta} \\ \hat{\lambda}^{-1}(1 - \hat{\beta})\hat{\alpha} & = \hat{\beta}_0 \\ \dfrac{1 - \hat{\beta}^2}{2\hat{\lambda}}\sigma^2 & = \sigma_{LS}^2 \end{cases}$$

Clearly, this system can be solved explicitly:

$$\lambda = -\frac{1}{h}\ln\hat{\beta}, \quad \hat{\alpha} = \frac{\hat{\lambda}}{1 - \hat{\beta}}\hat{\beta}_0, \quad \hat{\sigma} = \sqrt{\frac{2\lambda}{1 - \hat{\beta}^2}}\,\sigma_{LS}.$$

Problem 9.7 Consider quarterly prices $(S(t_1), S(t_2), S(t_3), S(t_4)) = (e^0, e^{0.3}, e^{0.5}, e^{0.4})$. Estimate parameters for a mean-reverting continuous time model.

Solution. Note that $h = 1/4$. Take $y_k = R(t_k) = \ln S(t_k)$, $x_k = y_{k-1}$, $k = 1, 2, 3$, and apply LS with $T = 3$ for the regression $y_t = \beta_0 + \beta y_{t-1} + \varepsilon_t$. Then the LS estimates of the parameters here are $\hat{\beta} = 0.2368$, $\hat{\beta}_0 = 0.3368$. Further, $\hat{\lambda}h = -\ln(\hat{\beta}) = 1.4404$, $\hat{\lambda} = 5.7614$,

$$\hat{\alpha} = \hat{\beta}_0\frac{\hat{\lambda}}{1 - e^{-\hat{\lambda}h}} = \hat{\beta}_0\frac{\hat{\lambda}}{1 - \hat{\beta}} = 2.5430.$$

We have fitted residuals

$$u_k = y_k - \hat{\beta}_0 - \hat{\beta} x_k, \qquad (u_1, u_2, u_3) = (0.0368, -0.0921, 0.0553),$$

hence the LS estimate for error variance (for a time series) is

$$\sigma_{LS} = \left[\frac{1}{T-1} \sum_{k=1}^{T} u_k^2 \right]^{1/2} = 0.0803,$$

hence

$$\hat{\sigma} = \sqrt{\frac{2\lambda}{1-\hat{\beta}^2}} \sigma_{LS} = 0.2805. \quad \square$$

Problem 9.8 Under the assumptions and notations of the previous problem, find a forecast of the return $R(t_4)$ and $R(t_5)$ based on observations for times (t_0, t_1, t_3). Presume that $t_k - t_{k+1}$ is three months for all k.

Solution. Let us use the estimates obtained in the previous solution. By (8.36), the desired forecast of $R(t_4)$ is $\mathbf{E}\{R(t_4)|y_0, y_1, y_2, y_3\} = \beta R(t_3) + \beta_0$. As was suggested in Section 8.9, we can replace this expectation by $\hat{\beta} R(t_3) + \hat{\beta}_0 = 0.2368 \cdot 0.4 + 0.3368 = 0.4315$. Similarly, (8.37) implies that the forecast of $R(t_5)$ is $\hat{\beta}^2 R(t_3) + (\hat{\beta} + 1)\hat{\beta}_0 = 0.2368^2 \cdot 0.4 + (0.2368 + 1) \cdot 0.3368 = 0.4390$. $\square$

Remark 9.9 It can be seen that the forecast above is different from the one obtained in Problem 9.6 for the same data set but for a different model. It is no surprise, since we have assumed more general autoregression now (in Problem 9.6, it was assumed a priori that $\beta = 1$). This outlines one more challenge for the reliability of forecasts for financial data: the results depend on the model choice. (See the discussion in Remark 8.33.)

9.4 Forecast of volatility with ARCH models

We assume that the risk-free interest rate r is a non-random constant. Consider the diffusion market model with a single stock such that its price $S(t)$ evolves as (9.1) and such that the volatility process $\sigma(t)$ is random. Typically, an equivalent risk-neutral measure is not unique in this case. If an equivalent risk-neutral measure is not unique on the σ-algebra generated by the stock prices, then the market cannot be complete (Theorem 5.35), i.e., there are claims $F(S(\cdot))$ that cannot be replicable. The question arises as to how to price these claims.

9.4.1 Black–Scholes formula and forecast of volatility square

For brevity, we shall denote by H_{BS} the corresponding Black–Scholes prices of different options, i.e., $H_{BS} = H_{BS,c}$ or $H_{BS} = H_{BS,p}$, for call and put respectively. Let

$$V \triangleq \frac{1}{T} \int_0^T \sigma(s)^2 ds.$$

Lemma 9.10 *Let V be non-random. Then:*

(i) the initial wealth $H_{BS}(S(0), K, V^{1/2}, T, r)$ ensures replication of the option claim;
(ii) $e^{-rT}\mathbf{E}_* F(S(T)) = H_{BS}(S(0), K, V^{1/2}, T, r)$ for any risk-neutral measure $\mathbf{P}_*$, where $\mathbf{E}_*$ is the corresponding expectation.[2]

Clearly, examples of a random volatility with non-random v can be invented: for instance, assume that $\sigma(t) = \sigma_1$ for $t \in [\tau, \tau + h)$, $\sigma(t) = \sigma_2$ for $t \notin [\tau, \tau + h)$, where $h > 0$, σ_1, σ_2 are given non-random values, and τ is a random time independent from $w(\cdot)$ and such that $0 \le \tau \le \tau + h \le T$ a.s. Then the corresponding V is non-random.

Proof of Lemma 9.10. Note that (ii) follows from (i). Therefore, it suffices to prove (i) only. Set $\tilde{K} \triangleq e^{-rT}K$, and set $f(x, K) = (x - K)^+$ or $f(x, K) = (K - x)^+$ for call and put respectively. We introduce the function $\hat{H}_{BS}(\cdot) : [0, T] \times \mathbf{R} \to \mathbf{R}$ such that

$$e^{rt}\hat{H}_{BS}(t, x) \equiv H_{BS}(t, e^{\rho(0)t}x, K, V^{1/2}, T, \rho(0)).$$

It is easy to see that

$$\frac{\partial \hat{H}_{BS}}{\partial t}(t, x) + \frac{x^2 V}{2} \frac{\partial^2 \hat{H}_{BS}}{\partial x^2}(t, x) = 0,$$

$$\hat{H}_{BS}(T, x) = f(x, \tilde{K}).$$

Let $\tilde{X}(t) \triangleq \tilde{H}_{BS}(\tau(t), \tilde{S}(t))$, where

$$\tau(t) \triangleq \frac{1}{V} \int_0^t \sigma(s)^2 ds, \quad \tilde{S}(t) \triangleq \exp\left(\int_0^t r(s)ds\right) S(t).$$

By Ito formula, we obtain that

$$d\tilde{X}(t) = \frac{\partial \hat{H}_{BS}}{\partial x}(\tau(t), \tilde{S}(t)) d\tilde{S}(t), \quad \tilde{X}(T) = f(\tilde{S}(T), \tilde{K}).$$

2 Hull and White (1987), p. 245.

Hence

$$\tilde{X}(0) = \hat{H}_{BS}(0, S(0)) = \mathbf{E}_*f(\tilde{S}(T), \tilde{K}) = e^{-rT}\mathbf{E}_*f(S(T), K).$$

This completes the proof. □

Corollary 9.11 *Assume that* $H_{BS} = H_{BS,c}$ *or* $H_{BS} = H_{BS,p}$. *Then*

$$e^{-rT}\mathbf{E}_*F(S(T)) = \mathbf{E}_*H_{BS}(S(0), K, V^{1/2}, T, r)$$

for any risk-neutral probability measure $\mathbf{P}_*$ *such that* $\sigma(\cdot)$ *does not depend on* $w(\cdot)$ *under* $\mathbf{P}_*$.

Proof. It suffices to observe that $e^{-rT}\mathbf{E}_*F(S(T)) = e^{-rT}\mathbf{E}_*\mathbf{E}_*\{F(S(T)) \mid V\}$. □

Clearly, $V = (1/T)\int_0^T \sigma(s)^2 ds$ is random in the general case of stochastic σ, and the assumptions of Lemma 9.10 are not satisfied. In this case, there is no simple solution for the option pricing problem (even if $\sigma(t)$ is constant over time).

By Lemma 9.10, it is natural to use an estimate (forecast)

$$\bar{V} = \frac{1}{T}\mathbf{E}_* \int_0^T \sigma(s)^2 ds \tag{9.13}$$

as a replacement for unknown and random V. For some volatility models, $\bar{V}$ can be estimated using well-developed ARCH and GARCH models for heteroscedastic time series. This approach may provide an approximate option price, when V is replaced by its estimate (forecast) $\bar{V}$.

9.4.2 Volatility forecast with GARCH and without mean-reverting

We consider again the stock price model (9.1)

$$dS(t) = a(t)S(t)dt + \sigma(t)S(t)dw(t).$$

Let $R(t) \triangleq \ln(S(t)/S(0))$. Assume that

$$R(0) = 0, \qquad dR(t) = \mu dt + \sigma(t)dw(t), \tag{9.14}$$

where μ is a constant, i.e.,

$$R(t) = R(t-h) + \mu h + \int_{t-h}^{t} \sigma(s)dw(s), \quad h > 0. \tag{9.15}$$

We assume also that $\sigma(t)$ with probability 1 is a piecewise constant function with jumps only in t_k, $k = 1, 2, 3, \ldots$, where $t_0 = 0$, $t_k = hk$, $h > 0$ is given. In addition, we assume that $\sigma(t) \equiv \sigma(t+0)$.

In this case, the process $R(t)$ is such that

$$R(t_k) = R(t_{k-1}) + \mu h + \sigma(t_k)\hat{\varepsilon}_{t_k}, \quad h > 0, \tag{9.16}$$

where $\hat{\varepsilon}_k = \int_{t_{k-1}}^{t_k} dw(s)$ are Gaussian random variables, $\mathrm{Var}\,\hat{\varepsilon}_k = h$. In other words,

$$R(t_k) = R(t_{k-1}) + \mu h + \varepsilon_k, \quad h > 0, \tag{9.17}$$

where

$$\varepsilon_k \triangleq q_k\eta_k, \quad q_k \triangleq \sigma(t_k)h^{1/2}, \quad \eta_k \triangleq h^{-1/2}\int_{t_k-h}^{t_k} dw(s).$$

Note that η_k are Gaussian random variables distributed as $N(0,1)$.

Let $T = Nh$, where N is an integer. We have that

$$\int_0^T \sigma(t)^2 dt = \sum_{k=0}^{N} \sigma(t_k)^2 h \sim \sum_{k=1}^{N} q_k^2. \tag{9.18}$$

Assume that we accept the hypothesis that the evolution of $q_k = \sigma(t_k)h^{1/2}$ is defined by the ARCH(1) model such that

$$q_{k+1}^2 = \alpha_0 + \alpha_1\varepsilon_k^2,$$

and that we have obtained estimates $(\hat{\mu}, \hat{\alpha}_0, \hat{\alpha})$ of $(\mu, \alpha_0, \alpha_1)$. In addition, assume that we estimate ε_0 as the corresponding residual, i.e., ε_0 be estimated as the corresponding residual, i.e., $\varepsilon_0 \sim R(t_0) - R(t_{-1}) - \hat{\mu}h$. By (9.18) and (8.49), we obtain

$$\mathbf{E}\int_0^T \sigma(t)^2 dt \sim \mathbf{E}\sum_{k=1}^{N} q_k^2 = \sum_{k=1}^{N}\left[\frac{\hat{\alpha}_0}{1-\hat{\alpha}_1} + \hat{\alpha}_1^{k+1}\left(\varepsilon_0^2 - \frac{\hat{\alpha}_0}{1-\hat{\alpha}_1}\right)\right]. \tag{9.19}$$

Assume that we accept the hypothesis that the evolution of q_k is defined by the GARCH(1,1) model such that

$$q_{k+1}^2 = \alpha_0 + \alpha_1\varepsilon_k^2 + \beta_1 q_k^2,$$

and that we have obtained estimates $(\hat{\mu}, \hat{\alpha}_0, \hat{\alpha}, \hat{\beta}_1)$ of $(\mu, \alpha_0, \alpha_1, \beta_1)$. Again, assume that we estimate ε_0 as the corresponding residual, i.e., ε_0 is estimated as the corresponding residual.

Let $T = Nh$, where N is an integer. By (9.18) and (8.56), we obtain

$$\mathbf{E}\int_0^T \sigma(t)^2 dt \sim \mathbf{E}\sum_{k=1}^{N} q_k^2$$
$$= \sum_{k=1}^{N}\left[\frac{\hat{\alpha}_0}{1-\hat{\alpha}_1-\hat{\beta}_1} + (\hat{\alpha}_1+\hat{\beta}_1)^{k+1}\left(\varepsilon_0^2 - \frac{\hat{\alpha}_0}{1-\hat{\alpha}_1-\hat{\beta}_1}\right)\right]. \tag{9.20}$$

9.4.3 Volatility forecast with GARCH and with mean-reverting

Assume that the process $R(t) \triangleq \ln(S(t)/S(0))$ evolves as

$$dR(t) = \mu(t)dt + \sigma(t)dw(t), \tag{9.21}$$

where

$$\mu(t) = \alpha - \lambda R(t),$$

where $\alpha \in \mathbf{R}$, $\lambda \in \mathbf{R}$ are unknown constants.

We assume again that $\sigma(t)$ with probability 1 is a piecewise constant function with jumps only in t_k, $k = 1, 2, 3, \ldots$, where $t_0 = 0$, $t_k = hk$, $h > 0$ is given, and that $\sigma(t) \equiv \sigma(t+0)$.

In this case, the process $R(t)$ is such that

$$R(t) = e^{-\lambda h}R(t-h) + \int_{t-h}^{t} e^{-\lambda(t-s)}\alpha ds + \int_{t-h}^{t} e^{-\lambda(t-s)}\sigma(s)dw(s), \quad h > 0. \tag{9.22}$$

Hence

$$R(t_k) = e^{-\lambda h}R(t_{k-1}) + \lambda^{-1}(1-e^{-\lambda h})\alpha + \sigma(t_k)\hat{\eta}_{t_k}, \quad h > 0, \tag{9.23}$$

where $\hat{\eta}_k = \int_{t_{k-1}}^{t_k} e^{-\lambda(t-s)}dw(s)$ are Gaussian random variables,

$$\mathbf{E}\hat{\eta}_k = 0, \quad \mathbf{E}\hat{\varepsilon}_k^2 = \hat{h}, \quad \hat{h} \triangleq \frac{1-e^{-2\lambda h}}{2\lambda}.$$

In other words,

$$R(t_k) = \beta R(t_{k-1}) + \beta_0 + \varepsilon_k, \tag{9.24}$$

where

$$\beta = e^{-\lambda h}, \quad \beta_0 = \frac{1 - e^{-\lambda h}}{\lambda}, \quad \varepsilon_k \stackrel{\Delta}{=} q_k \eta_k, \quad q_k \stackrel{\Delta}{=} \sigma(t_k)\tilde{h}^{1/2},$$

and where $\eta_k = \tilde{h}^{-1/2} \int_{t_k - h}^{t_k} e^{-\lambda(t-s)} dw(s)$ are Gaussian random variables distributed as $N(0, 1)$.

Assume again that we accept the hypothesis that the evolution of $q_k = \sigma(t_k)\tilde{h}^{1/2}$ is defined by the ARCH(1) model, $q_{k+1}^2 = \alpha_0 + \alpha_1 \varepsilon_k^2$, and that we have obtained an estimate $(\hat{\beta}, \hat{\beta}_0, \hat{\alpha}_0, \hat{\alpha})$ of $(\beta, \beta_0, \alpha_0, \alpha_1)$. In addition, we estimate ε_0 as the corresponding residual, i.e., $\varepsilon_0 \sim R(t_0) - \hat{\beta} R(t_{-1}) - \hat{\beta}_0 h$.

Let $T = Nh$, where N is an integer. Then (9.18) holds again. By (8.49), we obtain (9.19) again.

Assume that we accept the hypothesis that the evolution of q_k is defined by the GARCH(1,1) model $q_{k+1}^2 = \alpha_0 + \alpha_1 \varepsilon_k^2 + \beta_1 q_k^2$, and that we have obtained estimates $(\hat{\mu}, \hat{\alpha}_0, \hat{\alpha}, \hat{\beta}_1)$ of the coefficients. Again, assume that we estimate ε_0 as the corresponding residual. In this case, (9.18) holds again, with $T = Nh$, where N is an integer. By (8.56), we obtain (9.20).

9.5 Problems

Problem 9.12 Find $\mathbf{E}R_0(t)$ and $\mathrm{Var}\, R_0(t)$ for the mean-reverting model.

Problem 9.13 Find $\mathbf{E}\{R(t)|R(s)\}$ and conditional variance of $R(t)$ given $R(s)$, $s < t$, for a log-normal model without mean-reverting.

Problem 9.14 Assume that the stock price $S(t)$ is defined by Ito equation (9.1), (9.2). Find the conditional expectation $\mathbf{E}\{R(t)|R(s)\}$ under the assumption that $\mu(t) = \alpha - \lambda R(t)$, where $\alpha \in \mathbf{R}$, $\lambda \in \mathbf{R}$ are known constants, for $t = 2$, $s = 1$, $\alpha = 2$, $\lambda = 1$, $\sigma = 1$.

For the following problems, assume that all time periods $t_k - t_{k-1}$ are one-quarter of a year.

Problem 9.15 Assume that we are given the sequence of historical stock prices $(S(t_0), S(t_1), S(t_2), S(t_3)) = (e^0, e^{0.1}, e^{0.3}, e^{0.3})$, where $(t_0, t_1, t_2, t_3) =$ (January 1st, April 1st, July 1st, October 1st). Assume that the price $S(t)$ is defined by Ito equation (9.1), where $a \in \mathbf{R}$ and $\sigma \in \mathbf{R}$ are constants. Using the LS estimation procedure, estimate the parameters (a, σ).

Problem 9.16 Under the assumptions and notations of the previous problem, find a forecast of the return $R(t_4)$ and $R(t_5)$ based on observations for times (t_0, t_1, t_3) and presuming that $t_k - t_{k+1}$ is three months for all k.

Problem 9.17 Assume that the stock price $S(t)$ is defined by Ito equation (9.1), (9.2) such that $\mu(t) = \alpha - \lambda R(t)$, where $\alpha \in \mathbf{R}$, $\lambda \in \mathbf{R}$, $\lambda > 0$ (i.e., assume the mean-reverting model). Using the LS estimation procedure and the sample data $(S(t_k), t_k)$ given in Problem 9.15, estimate the parameters (α, λ).

Problem 9.18 Under the assumptions and notations of the previous problem, find a forecast of the return $R(t_4)$ and $R(t_5)$ based on observations for times (t_0, t_1, t_3) and presuming that $t_k - t_{k+1}$ is three months for all k.

Problem 9.19 Consider the following ARCH(1) model:

$$
\begin{aligned}
&y_k = \beta y_{k-1} + \varepsilon_k, \quad k = 0, 1, 2, \ldots, && \mathbf{E}\{\varepsilon_k \mid x_k\} = 0,\\
&\varepsilon_k = \eta_k \sigma_k \quad \text{with} \quad k = 0, 1, 2, \ldots, && \mathbf{E}_{k-1}\eta_k = 0, \quad \mathbf{E}_{k-1}\eta_k^2 = 1,\\
&\sigma_k^2 = \alpha_0 + \alpha_1 \varepsilon_{k-1}^2.
\end{aligned}
$$

In addition, assume that η_k are mutually independent and distributed as $N(0, 1)$, and η_k do not depend on $\{y_m\}_{m \leq k-1}$.

Here y_k are observable values, and α_0, α_1, and $a \in \mathbf{R}$ are unknown constants; $\mathbf{E}_k$ denotes the conditional expectation given observations $(y_k, y_{k-1}, y_{k-2}, \ldots)$.

Assume that we are given the sample of data $y_k = \ln S(t_k) - \ln S(t_{k-1})$, where $(S(t_k), t_k)$ are given in Problem 9.15. Using the ML approach, find out which set of estimates of the parameters $(\beta, \alpha_0, \alpha_1)$ is more likely to be the true one: $(\hat{\beta}, \hat{\alpha}_0, \hat{\alpha}_1) = (0.1, 0.5, 0.1)$ or $(\hat{\beta}, \hat{\alpha}_0, \hat{\alpha}_1) = (0.1, 0.3, 0.2)$.

Challenging problem

Problem 9.20 Suggest a method for estimating the parameters (σ_1, σ_2, p) under the hypothesis that the appreciation rate $a \in \mathbf{R}$ is a non-random constant, and the volatility $\sigma = \sigma(\omega)$ is random and has the two-point distribution described in Section 7.3.

Legend of notations and abbreviations

a.e. – almost everywhere, or for almost every
a.s. – almost surely
c.d.f. – cumulative distribution function
iff – if and only if
i.i.d. – independent identically distributed
p.d.f. – probability density function
$C(0,T)$ – the set of all continuous functions $f : [0,t] \to \mathbf{R}$
$\mathbf{E}\xi\eta = \mathbf{E}(\xi\eta)$
$F(S(\cdot))$ means that F is a function of the whole path $S(t)$, $t \in [0,T]$
$F(S.)$ means that F is a function of the whole path $S_1, \dots, S_T$
$\mathbb{I}_A$ is the indicator function of an event A
LS – least squares
ML – maximum likelihood
$N(a, \sigma^2)$ – the normal distribution law with mean a and variance σ^2
$\emptyset$ – empty set
2^Ω – the set of all subsets of Ω
$\sigma(\xi)$ is the σ-algebra of events generated by a random vector ξ
$x^+ \triangleq \max(x,0)$, $x^- = \max(-x,0)$
$|x| = \left(\sum_i x_i^2\right)^{1/2}$ for $x \in \mathbf{R}^n$
$x \triangleq X$ means that x is defined such that $x = X$
$\xi \sim N(0, v^2)$ – the random variable ξ has the distribution law $N(0, v^2)$
$\Leftrightarrow$ – if and only if
$\square$ – end of a proof or solution.

Selected answers and key figures

Chapter 1

Problem 1.58: $\mathbf{E}\{\xi \mid \eta\}(\omega_1) = 0$, $\mathbf{E}\{\xi \mid \eta\}(\omega_2) = 1/30$.

Chapter 2

Problem 2.33: (i) no, (ii) yes.

Chapter 3

Problem 3.67: $a < 1.07 < b$ or $a = b = 1.07$; Problem 3.69: 0.302.

Chapter 4

Problem 4.33: yes; Problem 4.67: The second expectation is $\left(\int_0^{5\wedge t} e^t dw(t)\right)^2 + \int_{5\wedge t}^5 e^{2t} dt$.

Chapter 5

Problem 5.71: $(a - r)^2 T$; Problem 5.74: $X(0) = e^{-3rT} e^{3\sigma^2 T} S(0)^{-2}$.

Chapter 6

Problem 6.31: 4.66; Problem 6.32: 5.78; Problem 6.33: yes in both cases.

Chapter 7

Problem 7.7: (c); Problem 7.10: 0.59; Problem 7.11: 0.7.

Chapter 8

Problem 8.18 (i): $(\beta_0, \beta) = (0.0833, 0.9792)$; Problem 8.38 (i): $(1.6333, -0.3333)$; Problem 8.39 (i): $(-0.3333)^k$.

Chapter 9

Problem 9.14: $0.36R(s) + 1.2642$; Problem 9.15: $(\hat{a}, \hat{\sigma}) = (0.42, 0.2)$; Problem 9.17: $(\hat{a}, \hat{\lambda}) = (0.82, 2.24)$; Problem 9.19: $(0.1, 0.3, 0.2)$.

Bibliography

[1] Avellaneda, M. (2000). *Quantitative Modeling of Derivative Securities: From Theory to Practice*. London: Chapman & Hall/CRC.

[2] Bachelier, L. (1900). Theorie de la speculation. *Ann. Ecole Norm. Sup.* **17**, 21–86. (English translation: Bachelier, L. Theory of speculation. In: *The Random Character of Stock Market Prices*, P.H. Cootner (ed.), MIT Press, Cambridge, Mass., 1967, pp. 17–78.

[3] Black, F. and Scholes, M. (1973). The pricing of options and corporate liabilities. *J. of Political Economics* **81**, 637–659.

[4] Cox, J., Ross, S., and Rubinstein, M. (1979). Option pricing: a simplified approach. *J. of Financial Economics* **7**, 229–263.

[5] Föllmer, H. and Schied, A. (2002). *Stochastic Finance: An Introduction in Discrete Time 2*. De Gruyter Studies in Mathematics, Berlin.

[6] Gujarati, D. (1995). *Basic Econometrics*. New York: McGraw-Hill.

[7] Higham, D.J. (2004). *An Introduction to Financial Option Valuation*. Cambridge: Cambridge University Press.

[8] Hull, J. and White, A. (1987). The pricing of options on assets with stochastic volatilities. *Journal of Finance* **42**, 281–300.

[9] Karatzas, I. and Shreve, S.E. (1998). *Methods of Mathematical Finance*. New York: Springer-Verlag.

[10] Korn, R. (2001). *Option Pricing and Portfolio Optimization: Modern Methods of Financial Math*. Providence, R.I.: American Mathematical Society.

[11] Lambertone, D. and Lapeyre, B. (1996). *Introduction to Stochastic Calculus Applied to Finance*. London: Chapman & Hall.

[12] Li, D. and Ng, W.L. (2000). Optimal portfolio selection: multiperiod mean-variance optimization. *Mathematical Finance* **10** (3), 387–406.

[13] Markowitz, H.M. (1959). *Portfolio Selection: Efficient Diversification of Investment*. New York: John Wiley & Sons.

[14] Neftci, S. (1996). *Mathematics of Financial Derivatives*. New York: Academic Press.

[15] Pliska, S. (1997). *Introduction to Mathematical Finance: Discrete Time Models*. Oxford: Blackwell.

[16] Shiryaev, A.N. (1999). *Essentials of Stochastic Finance. Facts, Models, Theory*. Hackensack, NJ: World Scientific.

[17] Söderlind, P. (2003). *Lecture Notes in Financial Econometrics*. University of St. Gallen and CEPR, Switzerland (web-published).

[18] Wilmott, P., Howison, S., and Dewynne, J. (1997). *The Mathematics of Financial Derivatives: A Student Introduction*. Cambridge: Cambridge University Press.

Index